NELSON SCIENCE



# *Force, Work, and Energy*

*Jim Wiese*

ITP Nelson Canada

I(T)P An International Thomson Publishing Company

Toronto • Albany • Bonn • Boston • Cincinnati • Detroit • London • Madrid • Melbourne • Mexico City
New York • Pacific Grove • Paris • San Francisco • Singapore • Tokyo • Washington

# Contents

 Important safety information

 A chance to use problem-solving skills

 Design or build

 A challenge

 Work together

 Record observations or data

# *Where Are the Forces?*

IMAGINE THAT YOUR CLASS is on a field trip to a local amusement park. You and your class are excited about going on the rides and all the fun you're going to have. But you have other plans for the day besides an exciting outing. You and your class are going to study the science that can be found in the rides and other activities at the amusement park. You have been split into groups, and each group has a different task. Your group's task is to investigate the forces that are found at the amusement park.

A force is a push or pull on an object and can be caused by a number of things, such as gravity, magnetism, or electricity. The force starts objects moving, or can cause them to stop moving, speed up, slow down, or change direction. The object that is pushed or pulled can even be yourself or another student.

The picture shows students at the amusement park. How many forces can you find?

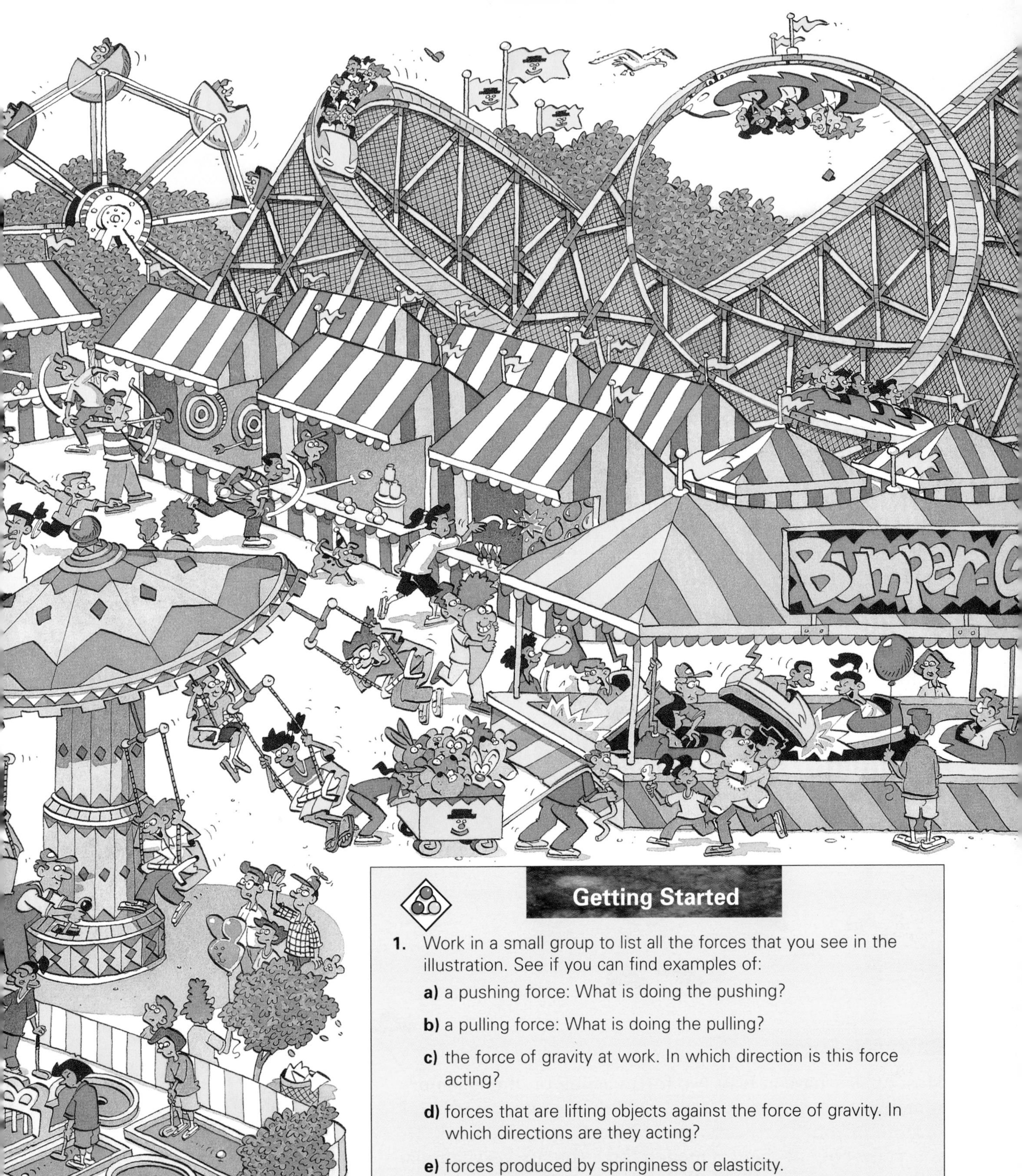


## Getting Started

1. Work in a small group to list all the forces that you see in the illustration. See if you can find examples of:
   **a)** a pushing force: What is doing the pushing?
   
   **b)** a pulling force: What is doing the pulling?
   
   **c)** the force of gravity at work. In which direction is this force acting?
   
   **d)** forces that are lifting objects against the force of gravity. In which directions are they acting?
   
   **e)** forces produced by springiness or elasticity.
   
   **f)** forces produced by things other than gravity.
2. What is the most common force that you find acting in the picture?
3. Look around the classroom. What forces can you see at work there?

2

# *Force*

YOU ENCOUNTER MANY FORCES every day, whether you are at an amusement park, or sitting in your classroom. Some of the forces you experience are the forces of gravity, friction, magnetism, electrostatics, and buoyancy.

**Forces** may push or pull things upwards, downwards, sideways, or in any direction. Forces can squeeze, bend, and twist things. To understand how a force acts on an object, you need to know in which direction a force is acting.

## Representing Forces

Forces can be represented by arrows. The way an arrow points shows the direction in which the force is acting. The length of the arrow represents the strength of the force. Look at the arrows in the photographs of the girl kicking the soccer ball. They show the direction and strength of the force acting on the ball.

## Balanced Forces

Most objects have at least two forces acting on them—and sometimes more. For example, hold this book in your hand in front of you. There are two forces acting on the book. Do you know both of them? One is a force you are very familiar with—gravity. It is pulling the book down. The other is the force applied by the muscles in your arm and hand, pushing up on the book.

When forces of equal strength act on an object in opposite directions, they are called **balanced forces**. For the book to move, either a new, third force must act on it, or one of the present two forces must become greater than the other.

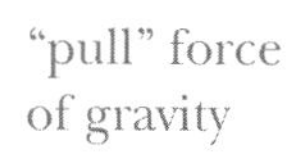


The two forces act in opposite directions and are equal, so the book does not move.

## Unbalanced Forces

Look at the students pulling the toboggan. The long arrow represents the force that moves the toboggan forward when the students pull the rope. The short arrow represents the force of friction acting on the toboggan. **Friction** is a force that resists motion. It acts to slow down the toboggan. These two forces, as the arrows show, are acting in opposite directions.

When one force acting on an object is greater than another force acting in the opposite direction, the forces are said to be **unbalanced**. Unbalanced forces will cause an object to move. The direction of motion of the object depends on which force is greater. To keep the toboggan moving forward, the pulling force on the toboggan must be greater than the force of friction.

### SELF CHECK

1. What is a force?
2. What is a balanced force? Give an example.
3. What is an unbalanced force? What does it cause? Give an example.
4. What is friction? Give an example.
5. Draw a person using a parachute. Label the forces acting on the parachutist. Are the forces balanced or unbalanced?
6. Draw a boat on a lake. Label the forces acting on the boat. Are the forces balanced or unbalanced?

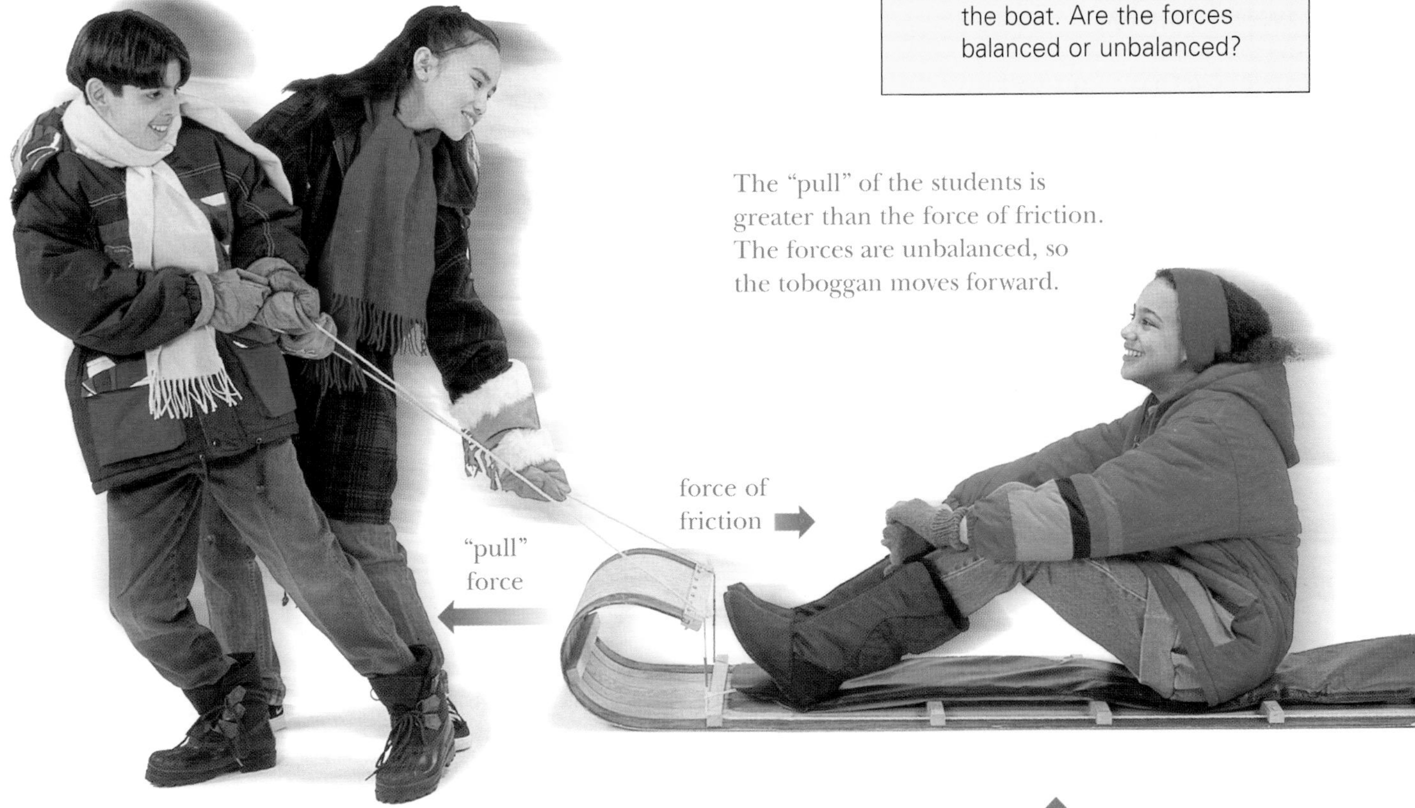


The "pull" of the students is greater than the force of friction. The forces are unbalanced, so the toboggan moves forward.

## Measuring Force

Using arrows to show both the strength and direction of all forces acting on an object will help you understand forces. But to know how strong a force is, you must measure it.

Force is measured in a unit called the **newton** (N). One newton is roughly the force required to support two golf balls, or one D-cell battery. The force of gravity on a 1 kg mass is approximately 10 N.

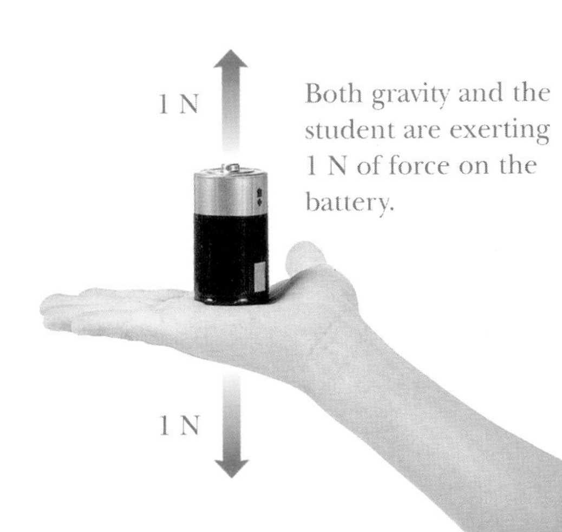


Both gravity and the student are exerting 1 N of force on the battery.

# Action/Reaction

ONE SCIENTIST WHO INVESTIGATED forces was Sir Isaac Newton. He summarized his theory about forces and their effects on motion. The summary is usually called Newton's Three Laws of Motion.

| Newton's Three Laws of Motion | |
|---|---|
| **Newton's First Law** | An object at rest will stay at rest and an object in motion will stay in motion, unless acted on by an outside force. |
| **Newton's Second Law** | An object will move with an acceleration— increasing speed—proportional to the force applied to it. |
| **Newton's Third Law** | For every action force, there is an equal and opposite reaction force. |

In this activity, you will investigate Newton's laws.

## Materials

- long balloon
- 4 m string
- drinking straw
- two chairs
- small rubber band
- tape
- scissors

## Procedure

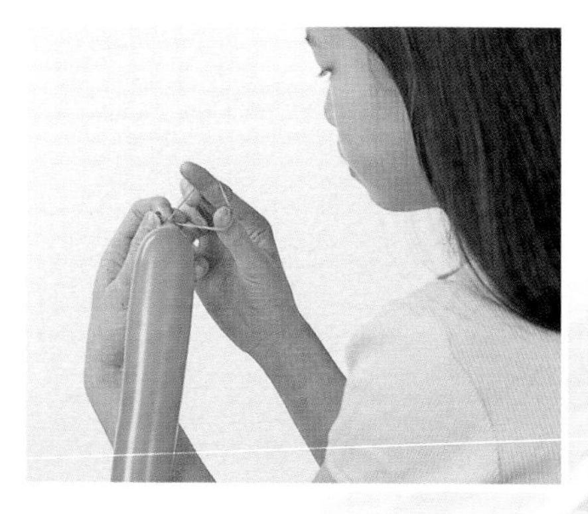

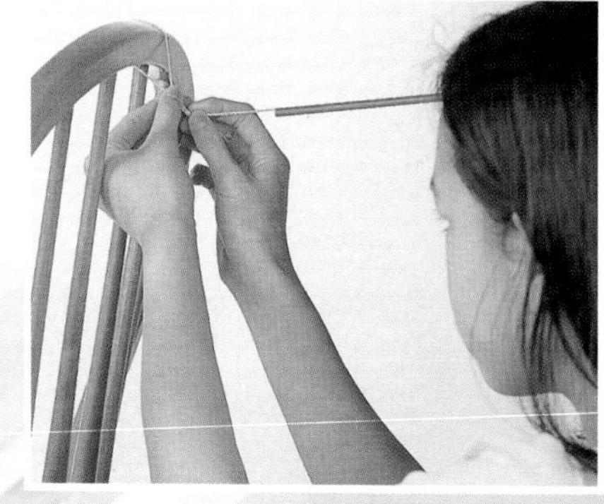

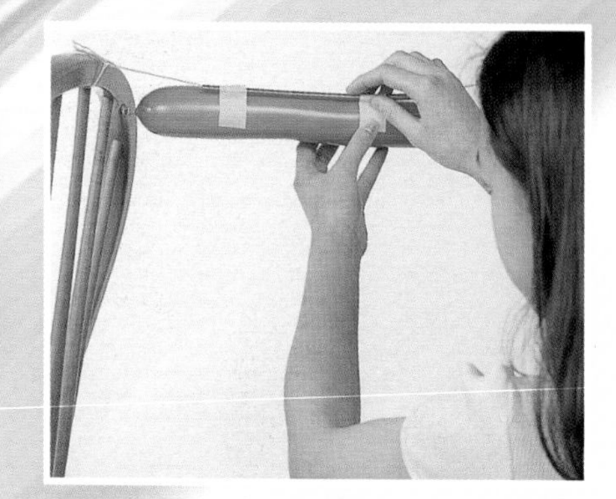

**1** Blow up the balloon. Twist the end closed, and use the rubber band to secure it.

**2** Thread the string through the straw.
■ Tie each end of the string to a chair. Separate the chairs so that the string is taut.

**3** Tape the balloon to the straw. Slide the balloon along the string until the tied end is next to one of the chairs.

a) Draw a diagram of your apparatus. What forces are acting on the balloon? Draw arrows to show the forces in your diagram.

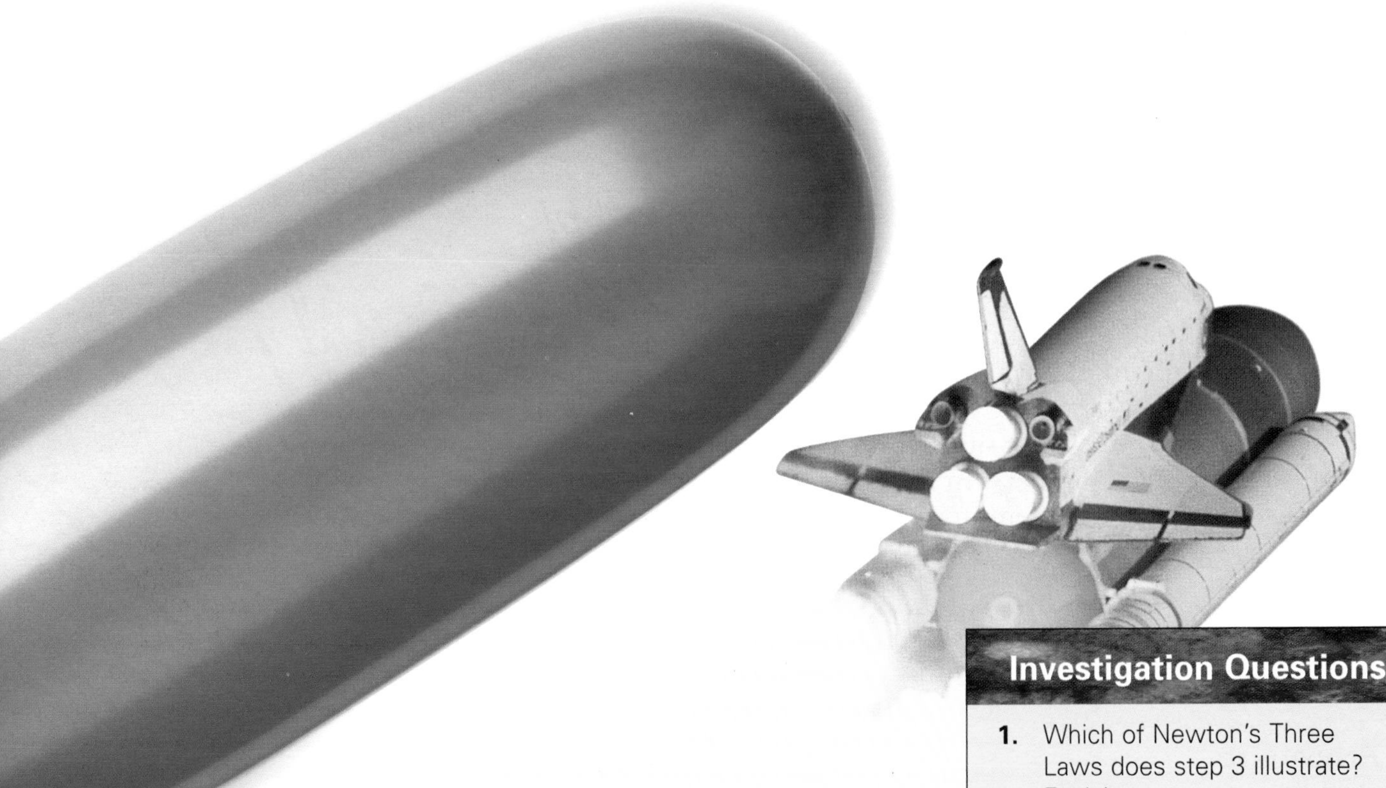

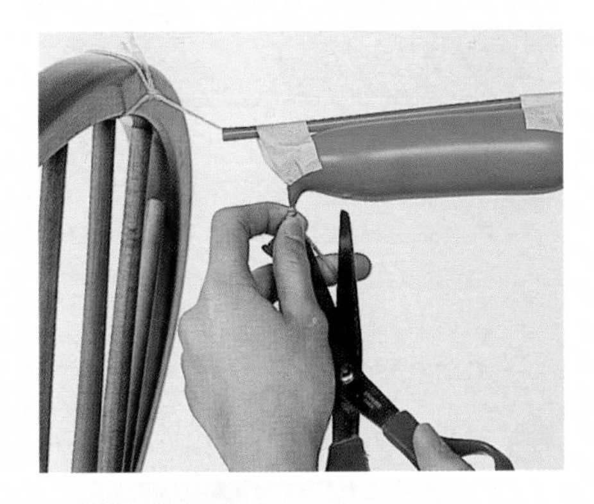

**4** Carefully cut the rubber band on the balloon and remove it, while holding the twisted end closed. ■ Release the twisted end of the balloon.

b) Draw a second diagram of your apparatus. What forces were acting on the balloon after you let it go? Draw arrows showing the forces in your diagram.

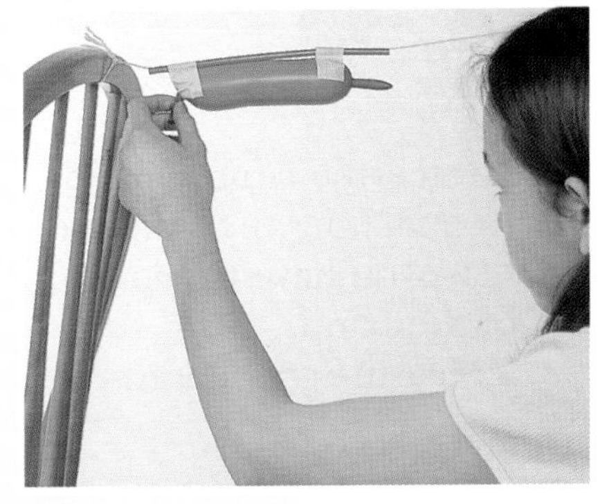

**5** Repeat the activity, but use less air in the balloon.

c) Record your observations.

## Investigation Questions

1. Which of Newton's Three Laws does step 3 illustrate? Explain.
2. Which of Newton's Three Laws does step 4 illustrate? Explain.
3. When the balloon moves down the string, what is the action force? What is the reaction force?
4. What forces cause the balloon to slow down and/or stop?
5. Which of Newton's Three Laws does step 5 illustrate? Explain.

## Apply

6. Suggest several places where you have seen Newton's Three Laws in action.

## Extension

7. Design an experiment to determine ways to make your balloon move farther or faster. Submit your experiment design to your teacher for approval, then do the experiment. Record your data and state your conclusion.

# Any Way You Stretch It

IF YOU TRY TO STRETCH, compress, or twist most materials, they will resist with an opposing elastic force. This is the kind of force you feel when you stretch a rubber band. Elastic forces are important to engineers designing everything from skyscrapers to cars. They're also important to bungie jumpers. In this activity, you will investigate the relationship between the length of a rubber band and its elastic force.

## Materials

- rubber bands
- pencil
- ruler
- various masses

## Procedure

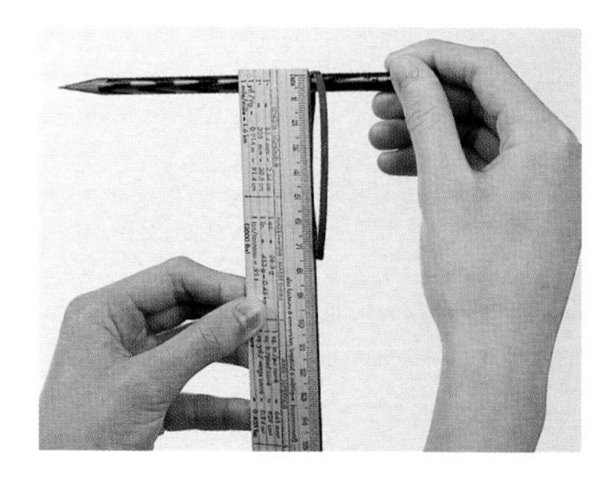

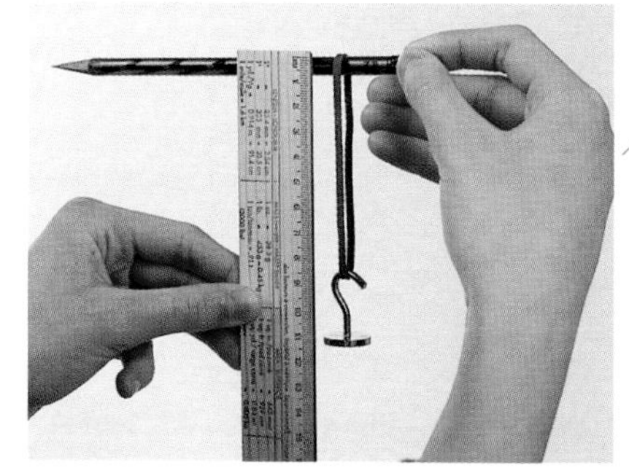

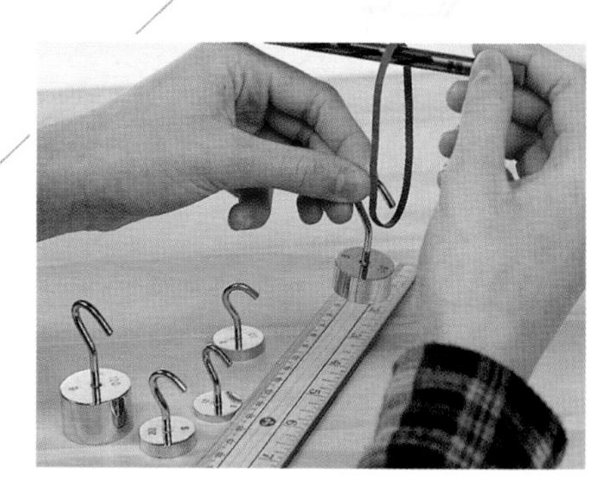

**1** Hook the rubber band around the pencil and hold it so the band is vertical. This is the natural length of the rubber band. Measure this length with the ruler.

a) Record the length of the band in a table like the one below.

| mass attached to rubber band (g) | length of rubber band (cm) |
|---|---|
| 0 | |
| 10 | |

**2** Stretch the rubber band by attaching the smallest mass you have to the end of the band.

- Measure the length of the band.

b) Record the length of the band in your table.

c) Draw a diagram to show the forces acting on the mass.

d) What is the balancing force that is opposing gravity?

**3** Gradually increase the mass attached to the band, measuring the length of the band with each increase.

e) Record the length of the band in your table.

f) What happens to the length of the band as you increase the mass?

**4** Using your data, plot a graph of mass against the length of the rubber band.

g) What does your graph look like? What does the shape of your graph mean?

h) What does the mass represent?

## Investigation Questions

1. Predict how long the rubber band would be after you added twice as much mass as you used in the investigation.
2. In an experiment, Sarah found that 100 g stretched her rubber band to 15 cm and 200 g stretched the band to 21 cm. What would be the length of her band if she added 150 g?

## Apply

3. What would your graph look like if you plotted mass against the increase in length of the band, instead of the total length?

   (Increase in length = total length – natural length)
4. What would the graph look like if a spring were used instead of a rubber band in this investigation?
5. Jas created the graph below.

   **a)** What was the natural length of his rubber band?

   **b)** How long would the band stretch if 100 g were attached to its end?

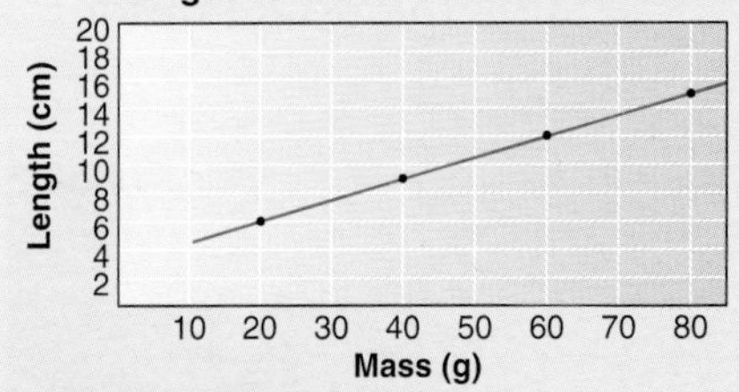


## Extension

6. Design an experiment to investigate the effects that length, thickness, or other factors have on the elastic force of the rubber band. Submit your design to your teacher for approval, then do the experiment. Record your data, create a graph, and state your conclusions.

# Making a Force Meter

HOW MUCH FORCE DOES IT TAKE to lift a brick or to pull a toboggan? We can use a ruler to measure length, and a clock to measure time. What kind of instrument could you use to measure the strength of forces? In this investigation, you will build such an instrument—a force meter.

## Materials

- wooden dowel
- masking tape
- hook
- plastic pipe
- rubber bands
- hose clamps
- various masses

## Procedure

*Making Your Meter*

**1** Place a strip of masking tape along one side of the dowel.
- Screw the hook into one end of the dowel.
- Slide the dowel inside the plastic pipe.

**2** Loop a rubber band around the hooked end of the dowel and tape it securely.

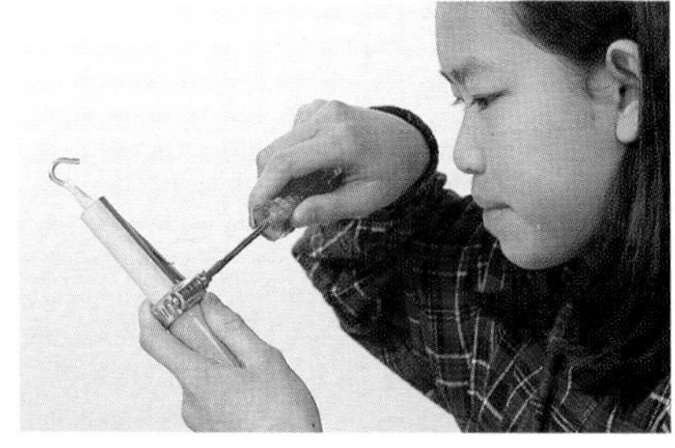

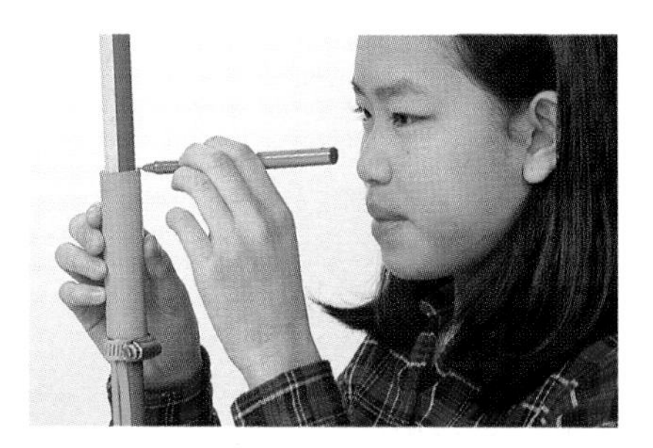

**3** Use the hose clamp to secure the other end of the rubber band to the plastic pipe.

**4** Hold the plastic pipe and turn the meter so the hooked end of the dowel is pointing down.
- Put a line on the tape at the edge of the pipe. Label the line "0."

*Calibrating Your Meter*

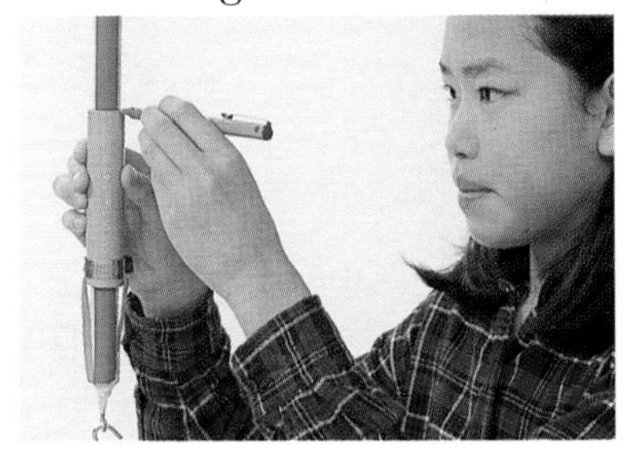

**5** Hold the meter up, with the hook end down, and place a 100-g mass on the hook.

- Put a line on the tape at the edge of the pipe. Label this line "1."
- Add another 100-g mass (the total is now 200 g) and draw another line. Label this line "2."
- Continue up to 1000 g (1 kg).

*Using Your Meter*

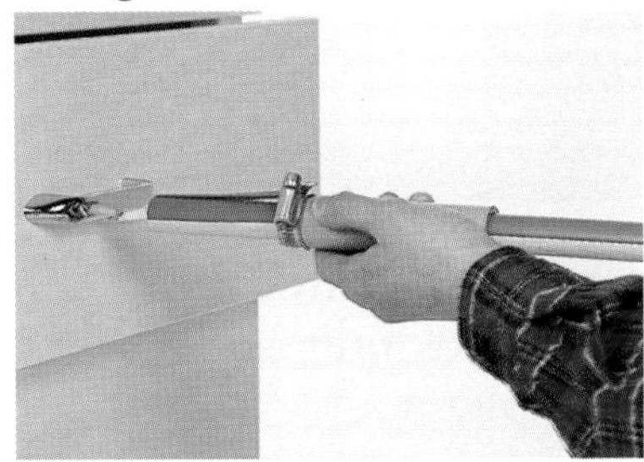

**6** First estimate, and then measure with your force meter, the force necessary to do the following:

a) Turn on a light switch.

b) Open a filing cabinet.

c) Open your locker.

d) Slide a textbook along your desk at an even speed.

## Investigation Questions

1. **a)** How close were your estimates in step 6? Which force reading surprised you the most?

   **b)** Draw a diagram of how you measured the force in each part of step 6. Use arrows to label the forces in each case.

2. What actions were difficult, or impossible, to measure using your force meter? Explain.

## Apply

3. Take your force meter home and measure the force necessary to do four or five actions, such as open a fridge door or turn on the television.

## Extension

4. Place a 100-g mass on your force meter and hold it vertically. Observe what happens as you make a small jump. Draw a diagram showing the forces at the top of the jump and when you land.

5. What improvements could you make to your force meter to make its measurements more accurate, or more easy to obtain? For example, what modifications would allow you to make any measurements you found difficult in step 6? Design another force meter that you think would also work or would work better. Submit your design to your teacher for approval, then make your force meter. Use the new meter to record data, just as you did earlier. State your conclusions.

# Friction

FRICTION IS A FORCE that resists motion whenever one material rubs against another. In this activity, you will investigate the effects of friction.

## Materials

- 2 books
- spring scale (or force meter with a scale in newtons–see Making a Force Meter)
- string
- piece of cardboard (30 cm × 50 cm)
- piece of wood (30 cm × 50 cm)
- piece of linoleum (30 cm × 50 cm)

## Procedure

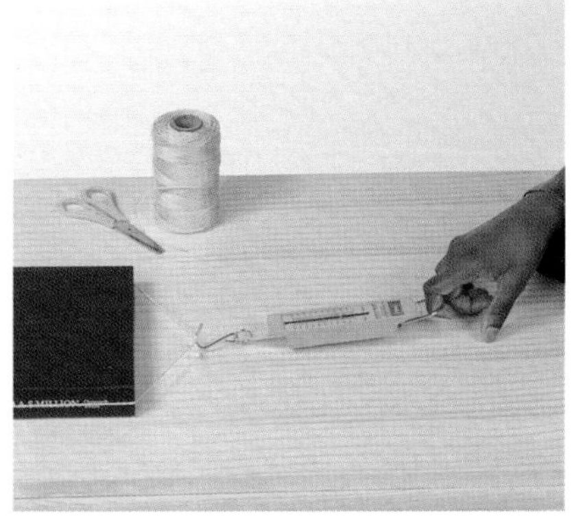

**1** Using the spring scale or force meter, pull the book across a wooden table with a steady pull.

a) Record how much force is necessary to keep the book moving.

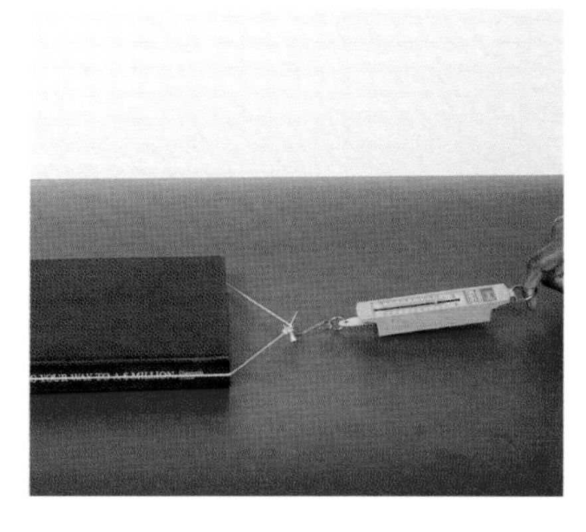

**2** Pull the book with the same steady pull across the linoleum.

b) Record the force.

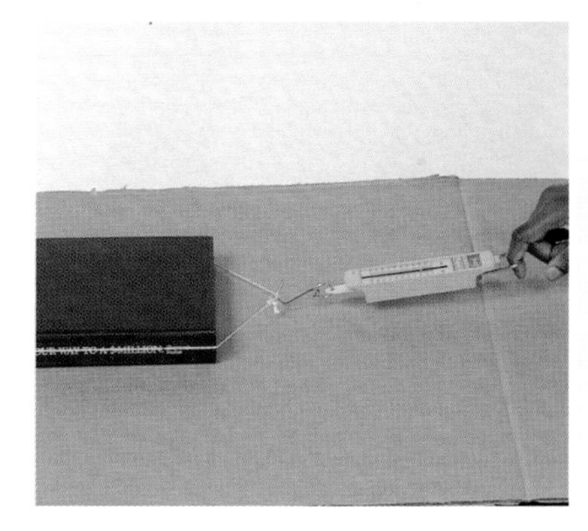

**3** Pull the book with the same steady pull across the cardboard.

c) Record the force.

Why do skiers wax their skis?

**4** Predict what will happen when you pull two books, instead of one, over each surface.

- Pull two books with a steady pull across each surface.

d) Write down your predictions.

e) Record the force necessary to pull the books across each surface.

f) Were your predictions correct? Explain.

## Investigation Questions

1. Compare the forces necessary to pull one book across the three surfaces. What is the effect of the surface on the amount of force needed to keep the book moving?
2. Draw a diagram of the moving book. Use arrows to show the forces acting on the book. Are the forces balanced or unbalanced?
3. What was the effect of doubling the number of books on the force needed to keep the books moving?

### Apply

4. In winter, drivers often put sandbags in the trunks of their cars. Do the results of your experiment explain why they do this?

### Extension

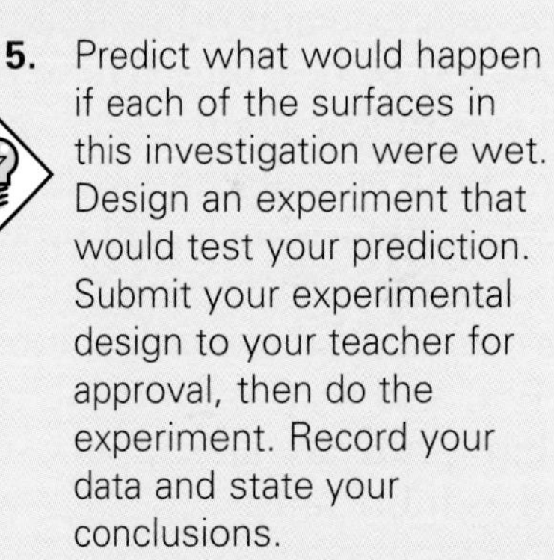

5. Predict what would happen if each of the surfaces in this investigation were wet. Design an experiment that would test your prediction. Submit your experimental design to your teacher for approval, then do the experiment. Record your data and state your conclusions.

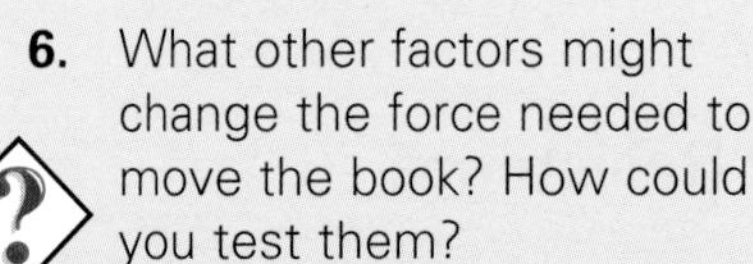

6. What other factors might change the force needed to move the book? How could you test them?

7. Is it easier to start an object moving, or to keep it moving once it has started to move? Write down your prediction. Design an experiment that would test your prediction. Submit your experimental design to your teacher for approval, then do the experiment. Record your data and state your conclusions.

# Overcoming Friction

WHEN YOU LOOK AT SURFACES CLOSELY, you can see that they are rough. Even ice looks rough if you magnify it enough. The rougher the surface, the more it resists movement of an object on it, and the stronger the force of friction.

You've seen that friction occurs when two objects move against each other. You've also noted that the strength of the force of friction will vary depending on the roughness or smoothness of the surfaces that come into contact with each other, and the mass of the moving objects.

Sometimes friction can be a lifesaver, when it slows down or stops a moving object. The brakes on a bike and in a car both rely on friction. But friction is not always useful. Wind resistance, a type of friction, and the friction between a bike's wheels and the axle make it more difficult for you to ride the bike.

## Reducing Friction

There are many ways to reduce unwanted friction. One way is to use bearings. Hard surfaces that move against each other in a machine are often protected by bearings. These bearings cause the surfaces to roll against each other, rather than slide. Both bicycles and skateboards use bearings between their wheels and their axles.

Another way to reduce friction is with lubricants. A **lubricant** is a substance that reduces friction and wear on moving parts. Oil, grease, and graphite are all slippery substances that are used as lubricants.

## Air Resistance

Have you ever tried to ride a bicycle into the wind? If you have, then you have felt the friction caused by air resistance. You might have tried to crouch down to decrease the resistance due to the air. Aerodynamics is the study of how the shape of an object affects the flow of air around it.

**Air resistance,** the friction of air, is so small at low speeds that we often do not notice it. At higher speeds, however, it becomes significant. One place where the friction of air is both a problem and a benefit is in the space shuttle.

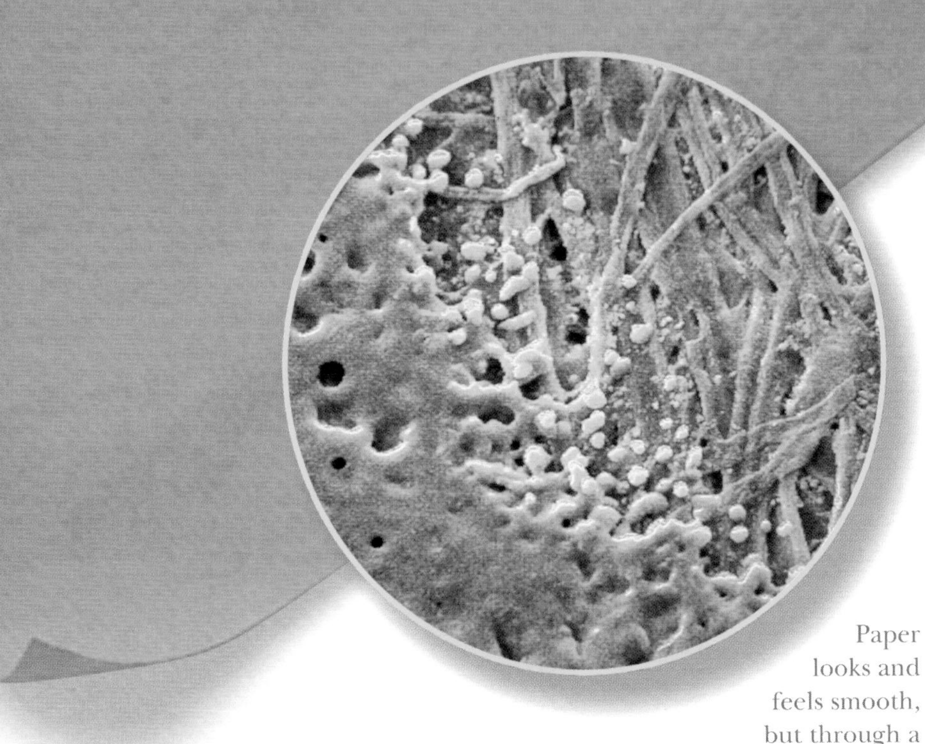

Paper looks and feels smooth, but through a microscope you can see that even the finest paper is rough enough to cause friction.

## SELF CHECK

**1. a)** List some ways, not included on these pages, that friction helps you in your everyday life.

**b)** List some ways, not included on these pages, that friction is a problem you have to overcome.

**2. a)** In the space shuttle, what problems does friction (air resistance) cause the spacecraft?

**b)** In the space shuttle, what benefits does friction provide?

### Extension

**3.** In your body, there are several places where surfaces rub together, causing friction. For example, bones meet at joints. How does your body reduce friction in your joints when you move? Research this subject, and present a report.

## Friction and the Shuttle

The space shuttle orbits 200 km above the Earth. At this distance, there is very little air, and so very little friction.

On return, the shuttle begins to hit more air molecules. The temperature of the hull begins to rise because of the friction between the air and the shuttle.

As the shuttle continues to fall, the friction between the air and the shuttle acts as a brake. As the speed drops, the friction lessens.

When the shuttle finally reaches Earth, three parachutes are released to act as air brakes, further slowing the shuttle.

# *Buoyant Force*

ANOTHER FORCE THAT WE SEE AROUND us is the buoyant force. Buoyancy is the upward force of water that works against an object's mass when the object is in the water. In this activity, you will investigate how the buoyant force works to make some objects float.

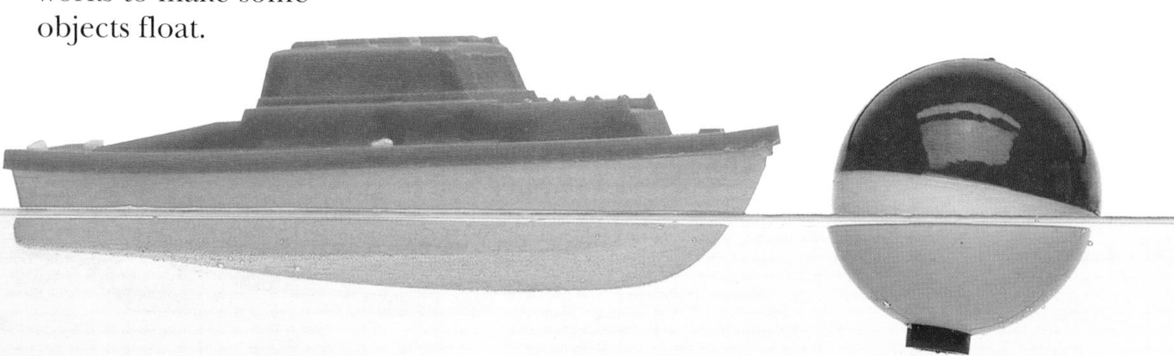

## Materials

- styrofoam cup
- tub of water
- metal washers
- balance scale

## Procedure

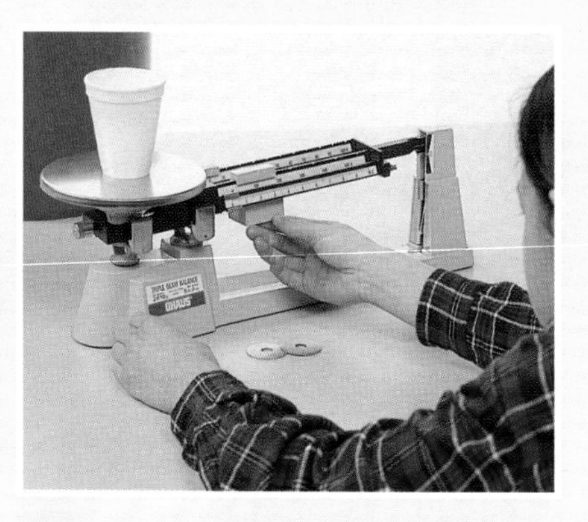

**1** Place the styrofoam cup upright in the water.

a) Predict how many washers you will be able to place in the cup before it sinks.

**2** Begin adding washers to the cup. Continue adding washers until the top of the cup is even with the water level.

**3** Remove the cup with the washers from the water, dry it, and measure the mass. Count the number of washers in the cup.

b) Record the mass of the cup with washers.

c) Record the number of washers. How accurate was your prediction?

## Investigation Questions

1. Compare the mass of the cup with the washers and the mass of the cup filled with water. Why do you think you obtained those results?
2. If you drew a line halfway up the cup and added washers until the water level reached that mark, how much water would you add to the cup so that its mass was the same as the mass of washers?
3. Which of the following would float: a kilogram of wood or a kilogram of steel?

### Apply

4. Use the results of this experiment to explain how a wooden rowboat floats.
5. Steel will not float, yet a large metal ship will stay on top of the water. Explain.

### Extension

6. Use a 20-by-20 cm square of aluminum foil to build a boat that will hold the most washers without sinking. Place metal washers in the boat until it sinks. What design of boat holds the most mass without sinking? Do ships use that design? Explain.

7. Compare the shape of a supertanker and a racing sailboat. Explain how the design of each matches its purpose.

**4** Take the washers from the cup and fill the cup to the top with water. Weigh the cup filled with water.

d) Record the mass of the cup full of water.

# *Building Forces*

ENGINEERS USE A KNOWLEDGE OF FORCES every day. Before the CN Tower was built, for example, engineers had to study all the forces that would be acting on the tower so their design would be strong enough to stand up under even the worst conditions, such as wind, snow, or even an earthquake.

Jamil Mardukhi was one of the engineers who helped build the CN Tower. In 1973, when he was still a student at the University of Toronto, he started as site engineer, supervising the construction of the tower. Now he's the engineer in charge of the tower structure, and a partner in Nicolet Chartrand Knoll, the company of engineers that designed the tower.

As site engineer, he made sure that all of the work was done according to the design drawings. He says the job was very exciting, and very demanding. "We were using new construction techniques. Concrete was being poured 24 hours a day, which never stopped except once on weekends for cleaning."

A lot of engineering work involves computers and computer drawing. For this career, and many others, knowledge of math, science, and computers is very important.

One part of an engineer's job is designing new structures, like a new building, a hydroelectric dam, a bridge, or even another tower. If you like to design and build things, try the activity on the facing page. It will help you better understand how engineers do their work. If you enjoy doing this type of activity, you could consider becoming an engineer.

From design to the capping of the CN Tower with its antenna, engineers had to deal with many forces.

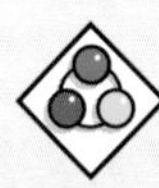

**TRY THIS**

### Straw Tower

Engineers do many different jobs. One job is to plan and design structures. In this activity, you are going to plan, design, and then build a tower.

Your group's task is to build the tallest tower possible using only 50 plastic drinking straws, 1 m of masking tape, and scissors.

- If possible, do some research before you begin. Look at towers in your community, or look for tower structures in books. What do they look like? Why do you think they have the shape they do?
- Draw a sketch of what your tower will look like.
- Use the straws and masking tape to build your design.
- Experiment! Try other designs; see if you can do better each time.

**a)** Look at the tallest towers that were built in your class. What characteristics, if any, do the tallest towers share?

**b)** What was the biggest challenge that you faced in building your tower higher?

**c)** If you were to build another tower, what would you do differently?

## Extension

1. Build another tower, only this time don't build it for maximum height. Build it to be 75 cm tall and as strong as possible to withstand wind. Design a way to test its ability to stand up to wind and snow storms.
2. Build another tower designed to withstand an earthquake. How is the design of this tower different from the others?

# *Building for Earthquakes*

HAVE YOU EVER WONDERED if your town will have an earthquake? And if it did have an earthquake, could the buildings withstand the forces it would create?

If you lived in Kobe, Japan, you'd know the answer. On January 17, 1995, an earthquake of magnitude 7.2 hit the city of Kobe. Although the tremor lasted only 20 seconds, the damage to the city was enormous, leaving 5502 people dead and $99 billion in property damage.

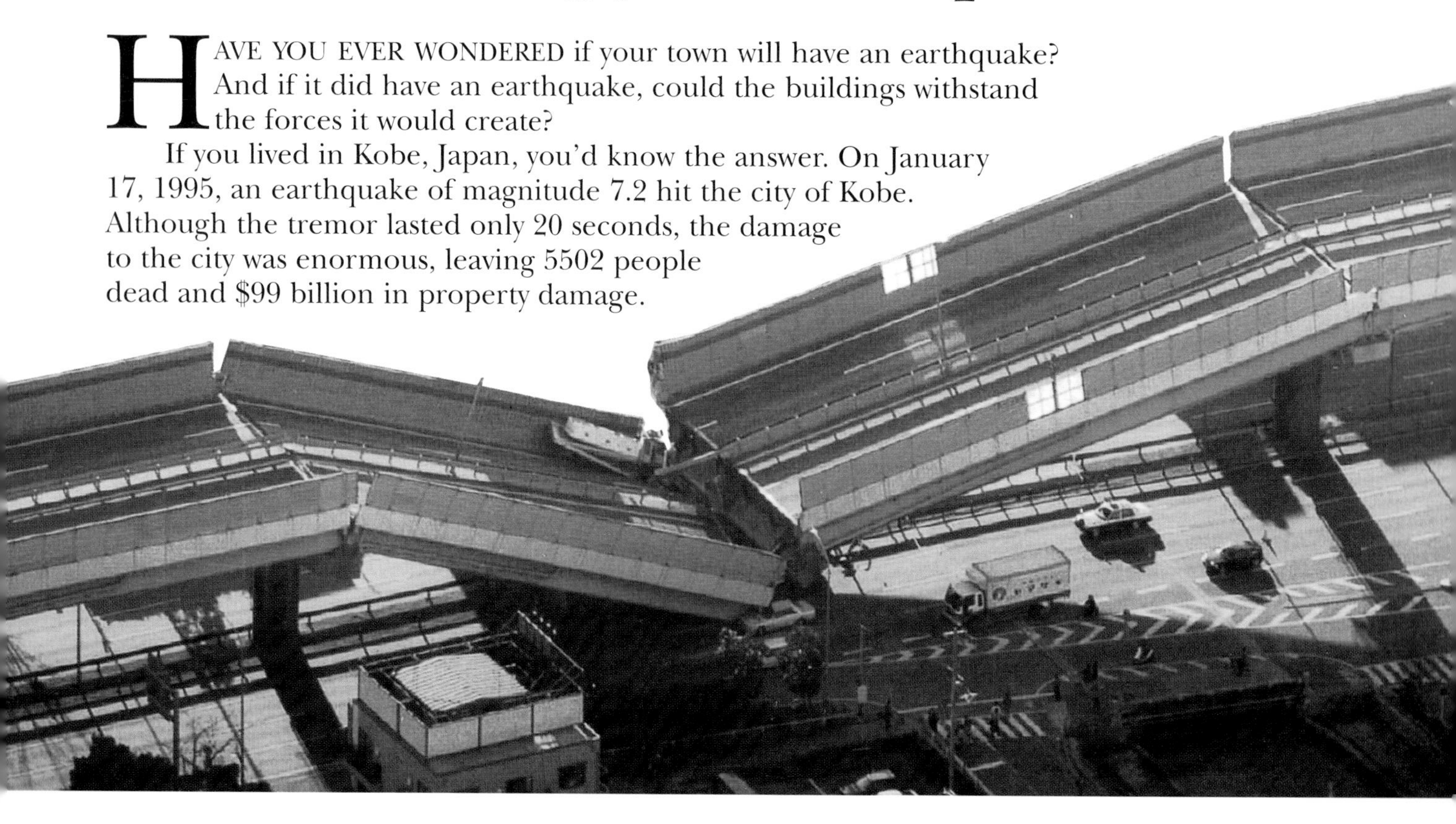

## How Earthquakes Happen

An earthquake is caused by the fracture or sliding of rock within the Earth's crust. Fifteen 100-km-thick plates of rock make up the Earth's crust. These plates are moving very slowly, and they continually collide and slide past each other. Earthquakes occur mostly at the boundaries of these plates.

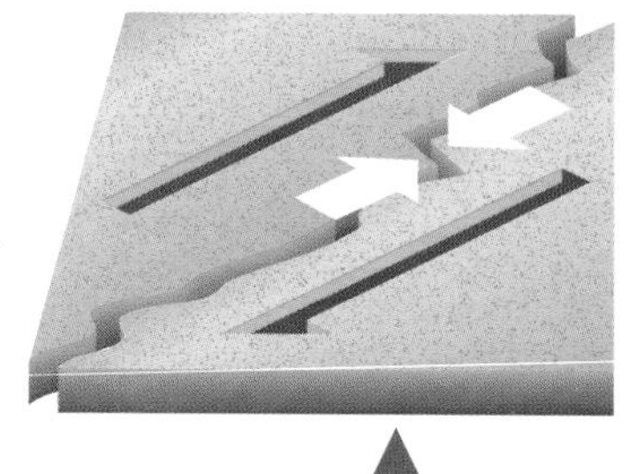

The surfaces of the plates are rough. Friction prevents the plates from moving and pressure builds up until there is a sudden "jump," releasing massive amounts of energy.

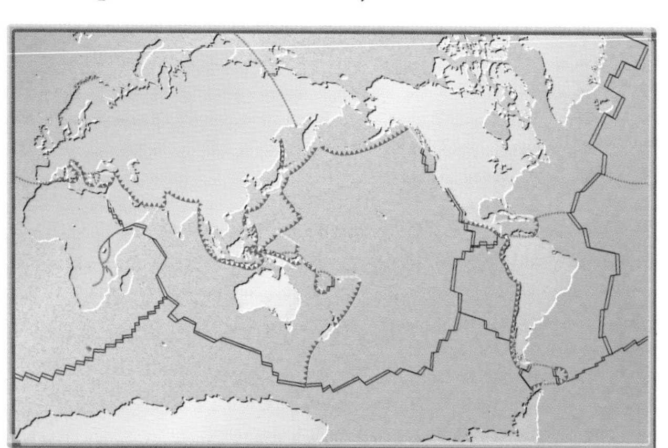

Huge plates move across the Earth, carrying the continents. Where the plates meet, earthquakes are more likely.

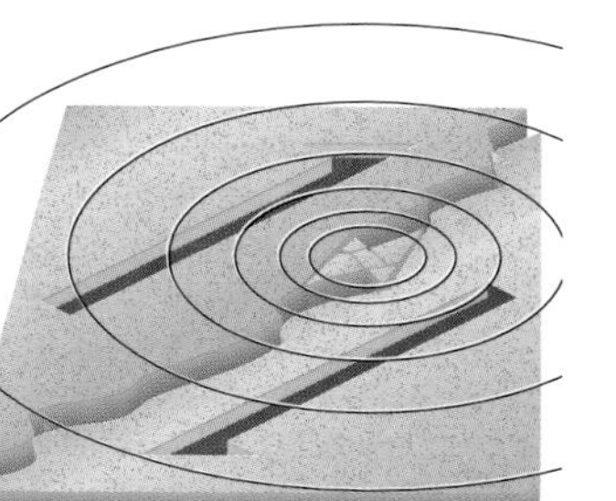

The energy spreads through the rock in waves, causing earthquakes on the surface.

## Earthquakes in Eastern Canada

Eastern Canada is in a stable region of the North American Plate and has a relatively low rate of earthquake activity. Almost all of the quakes in this region (99%) measure 4 or less on the Richter scale and are barely felt by people. (The Richter scale is a way to measure the size of an earthquake. An earthquake measuring 5 on the Richter scale will have 10 times the energy of a quake measuring 4. A quake that measures 6 will have 100 times the energy of a quake measuring 4.)

Although most earthquakes in Eastern Canada are small, over an average 30-year period, engineers would expect to encounter three quakes of more than 5 on the Richter scale. An earthquake that size can damage buildings. When an earthquake hit Mexico City in 1985, most of the deaths were caused by collapsing buildings.

## Preventing Damage

Fortunately, there are ways to minimize damage from an earthquake. The most effective way is through building safer and stronger structures, but there is a cost. In regions where earthquakes are uncommon, engineers usually design buildings to withstand small earthquakes. They can design a building to withstand larger earthquakes. But the construction costs are higher, sometimes double.

This building was designed to be "earthquake-proof".

### SELF CHECK

1. Where do most earthquakes occur?
2. What is the Richter scale used to measure? How does it work?
3. Do you think earthquakes are a concern where you live?

### TRY THIS

Debate

### Proposal

All buildings within our town must be built or modified to be able to withstand an earthquake measuring 6 on the Richter scale. Schools and hospitals must be able to withstand an earthquake measuring 7.

### Point

- Improved construction standards will make buildings safer in an earthquake.
- Most people who are killed in an earthquake die due to collapsing structures.
- The technology exists to make all buildings safer.

### Counterpoint

- Current building standards are strict enough. The buildings in our town have stood up for many years and will stand for many more.
- Higher construction standards will make buildings cost more to build. Who is going to pay the increased cost?
- No one can predict when a big earthquake will hit. Maybe we won't have one in the next 100 years.

### What Do You Think?

Research the issue further, expand upon the points provided, and develop or reflect upon your position. Prepare for a class debate.

# Work

AS YOU HAVE LEARNED, forces can act on an object. Lift this book off the desk and into the air. In science, if a force is exerted on an object and the object moves for a distance in the direction of the force, then **work** has been done. By lifting the book, you have done work.

Now push down on your desktop. Does it move? No? Then you didn't do any work, no matter how hard you pushed.

Hassan has been pulling, pushing, and heaving on this bag for 15 minutes, but it hasn't moved. He has not done any work on the bag.

Hassan and Andrea have lifted the bag from the ground onto the wheelbarrow. They have done work on the bag, because they have exerted a force on it (equal to the force of gravity on the bag), and have also lifted the bag a certain distance (equal to the height of the wheelbarrow).

## Calculating Work

When a force moves an object, the amount of work done on the object depends on two factors:

- the amount of force exerted
- the distance the object moves

These two factors are used to calculate the amount of work done on an object, as shown in the following equation:

$$\text{work} = \text{force} \times \text{distance}$$

This same equation can be written using symbols:

$$W = F \times d$$

## Work Units

When using the equation for calculating work, it is important that you use the correct units of measurement. Distance is measured in metres (m), and force is measured in newtons (N). The measuring unit for work is the newton metre (N•m). The newton metre has been given the special name **joule** (J). It is named after James Joule (1818–1889), a physicist who did many experiments on work and energy. One joule is the amount of work done when a force of one newton pushes or pulls an object a distance of one metre (1 J = 1 N•m). Notice that the unit of work (the joule) is the same as the unit of energy. This is because work is energy that has been transferred from one object to another.

The example below will help you learn how to use the equation to calculate work.

## Example

To raise a box of books from the floor to a shelf, Cheryl exerts a force of 200 N. The box moves a distance of 1 m. How much work did Cheryl do on the box?

## Solution

1. Write down the work equation.

   $W = F \times d$

2. Substitute your values for force and distance. Be sure to include the units. In this case, the force is 200 N, and the distance is 1 m.

   $W = 200\ \text{N} \times 1\ \text{m}$

3. Multiply the numbers.

   $W = 200\ \text{N•m}$

4. State your answer in joules. (Remember, 1 J = 1 N•m.)

   $W = 200\ \text{J}$

5. Write your answer in a sentence.

   *Cheryl did 200 J of work on the box lifting it 1 m to the shelf.*

In the example above, the work Cheryl does is used to push upward against the force of gravity on the box of books. If Cheryl used the force to push the box along the floor for some distance, she would still be doing work. However, the work would not be done to overcome gravity; it would be done to overcome friction.

### SELF CHECK

1. What is work?
2. In science, all words have precise meanings. Because we use many of the same words in everyday language, it may seem strange to use these words in a specific way. Identify the following meanings for the words energy, work, and force as either scientific usage or everyday usage. Explain your answers.
   **a)** She forced me to eat the last piece of pizza.
   **b)** I had to pull my brother in his wagon all around the block. What a lot of work!
   **c)** I was amazed to see how little force was necessary to get the ice sled moving.
   **d)** I don't have the energy to concentrate on that problem.
3. Laurie needs a force of 40 N to pull her younger brother on a toboggan. If she pulls him a distance of 150 m, how much work has she done?

# Simple Machines

A **MACHINE** IS A DEVICE THAT HELPS PEOPLE do work more easily. For example, to lift heavy furniture straight up onto the back of a truck, you would probably need several friends to help. With a ramp, however, you can do this work more easily.

## Machine Functions

Every machine performs at least one of these functions:

- A machine may change the direction of a force.

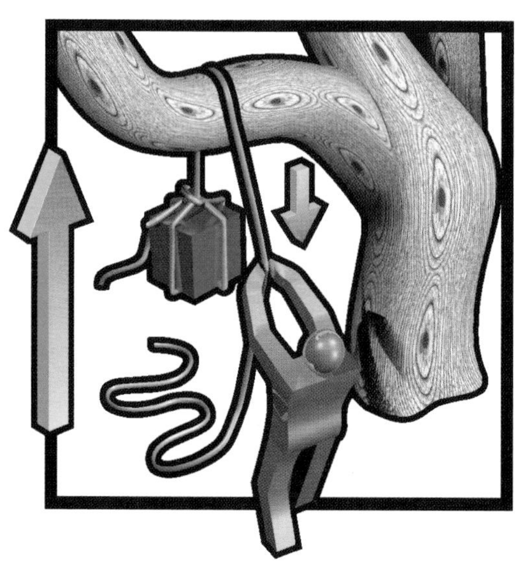

A rope thrown over a tree branch can be used to lift a box. The box is lifted up by pulling down on the opposite end of the rope.

- A machine may multiply speed or distance.

The combination of the bicycle's wheel and axle with pedals and chain multiplies speed.

- A machine may transfer forces from one place to another.

The chain of a bicycle transfers force from the pedals to the rear wheel.

- A machine may multiply force.

A small force applied to the handle of the jack results in a force large enough to lift a car.

## Six Simple Machines

All machines, no matter how complex, are made up of one or more of six **simple machines.**

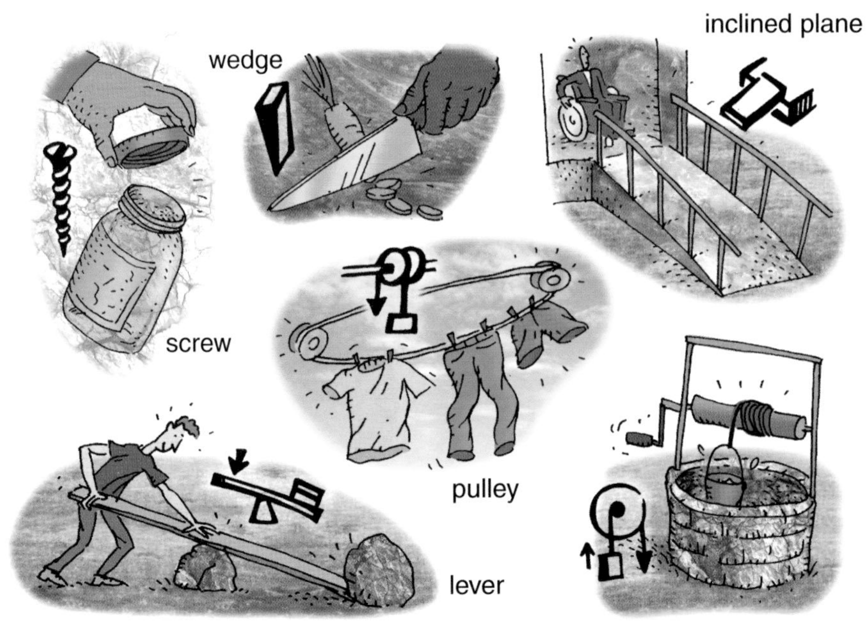


## Mechanical Advantage

The number of times that a machine multiplies a force is called the machine's mechanical advantage. You can calculate the **mechanical advantage** of any machine by using the following equation:

$$\text{mechanical advantage} = \frac{\text{load force}}{\text{effort force}}$$

**Load force** is the force needed to move the object without the machine. **Effort force** is the force needed to move the object with the machine.

## Example

A boy uses an effort force of 100 N to move up a ramp. If he had to lift himself straight up against gravity, the load force would be 900 N. Calculate the mechanical advantage.

$$\text{mechanical advantage} = \frac{\text{load force}}{\text{effort force}}$$

$$\text{mechanical advantage} = \frac{900\ \text{N}}{100\ \text{N}}$$

$$= 9$$

This means that the ramp multiplies the effort force by 9 times. To do the same amount of work, the boy uses 1/9th of the force needed to lift himself straight up, but has to move 9 times as far along the ramp.

## SELF CHECK

1. What functions can a machine perform?
2. In the middle of a sheet of paper, write "Simple Machines." Around this phrase, write the names of the six simple machines, then create a mind map using the ideas you've learned. Add everyday examples of machines.
3. Maria uses a lever to pry open a lid of a paint can. The load force is 100 N and the effort force is 20 N.
   **a)** What is the mechanical advantage of the lever?
   **b)** Is Maria using the lever to multiply force or distance?
4. Suppose you use your forearm to lift a ball.
   **a)** What kind of simple machine is your forearm?
   **b)** What is the machine function of your forearm?
5. The huge pyramids in Egypt were constructed with the help of simple machines. Large stone blocks used probably came from far away. Describe what simple machines might have been used to cut stone blocks, take them out of the ground, move them, and raise them to the layers of the pyramids.

# Working Your Way Up The Ramp

THERE ARE SEVERAL WAYS THAT YOU CAN use simple machines to help make your work easier. In this investigation, you will study two simple machines. You'll learn more about simple machines later in this unit.

## Materials

- block of wood with a hook in it
- 5-N spring scale
- smooth board for a ramp
- several books
- support stand with ring
- single pulley
- string (1.5 m)

## Procedure

**1** Use the spring scale to lift the block straight up. The reading on the scale shows the force of gravity being exerted on the block.

a) Record the force reading on your scale.

**2** Use books and a board to make a ramp tilted at about 30°.

- Use the spring scale to pull the block up the ramp at a steady speed.

b) Record the force reading on the scale.

**3** Connect one end of the string to the ring on the support stand, pass it through the pulley, and connect the other end to your spring scale.

- Connect the pulley to the block.
- Use the pulley to lift the block.

c) Record the force reading on your scale.

4 Use the pulley to pull the block up the ramp.

d) Record the force reading on your scale.

Thousands of years ago, people used ramps, pulleys, and other simple machines to move massive pieces of stone. We still do.

## Investigation Questions

1. How did the force needed to move the block in steps 2, 3, and 4 compare with each other?
2. How much did the ramp-and-pulley system decrease the force needed to move the block: by about one-quarter, one-third, or one-half?
3. What were the advantages and disadvantages of using these simple machines to help move the block?
4. What other simple machines could have been used to move the block with less force?

### Apply

5. Try to improve the system using the materials provided. Is it possible to decrease the force necessary to move the block even more?
6. List examples where you have seen ramps and pulleys used to move heavy loads.

### Extension

7. Devise a system to lift a given load to a certain height, using the least force possible. Design an experiment that would test your system. Submit your experimental design to your teacher for approval, then do the experiment. Record your data and state your conclusions.

# A Not-So-Simple Machine

MOST OF THE MACHINES WE ENCOUNTER in our lives are really not so simple. Many machines contain several simple machines. In this investigation, you will examine the simple machines found in a bicycle.

## Materials

- one bicycle

## Procedure

Examine the bicycle, and answer the following questions.

Your answers to these questions may depend on the type of bicycle you examine. Write down the type of bike used—ten-speed, five-speed, BMX bike, mountain bike,…

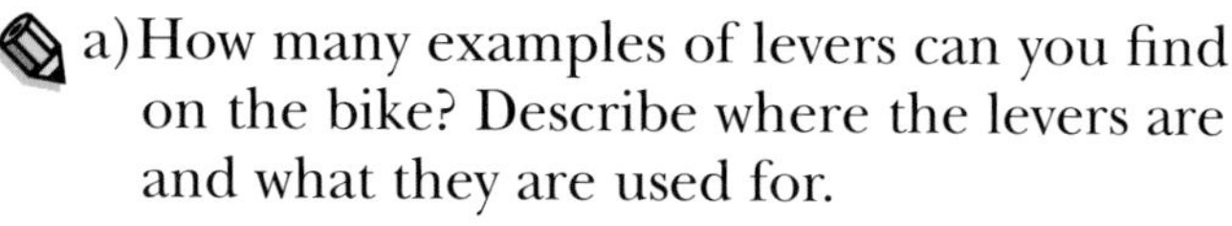

a) How many examples of levers can you find on the bike? Describe where the levers are and what they are used for.

b) What parts of the bike would you identify as examples of a screw? What is each part used for?

c) What parts of the bike are examples of a wheel-and-axle? What is each used for?

d) Can you find examples of a pulley, an inclined plane, and a wedge on the bike? What is each used for?

## Investigation Questions

1. What do you think the purpose of the gear sprockets is?

2. Describe how the brakes work on the bike. What force do they create?

### Apply

3. Examine several different types of bikes (BMX, ten-speed, one-speed,...). What special features does each have that make it suitable for its specific use?

4. **a)** Design, on paper, a mechanical device that uses at least three kinds of simple machines (inclined plane, pulley, lever, wedge, wheel-and-axle, and screw). Your invention must perform a useful function. Some suggestions for your invention are:
   - chalkboard cleaner
   - fence-painting machine
   - wake-up alarm
   - window opener
   - burglar alarm

   Don't feel limited by this list—go ahead and use your own ideas.

   **b)** Build a model of your device using Lego, Meccano, or other materials you can find around your house.

# Galileo and Aristotle

MANY SCIENTISTS and philosophers have studied the force of gravity. As far as we know, the first to write down what he thought about why objects fall to earth was Aristotle, around 350 B.C. He felt that all objects had a "natural place" and when removed from their natural place, they tended to return. So fire and smoke rise through the air to return to their natural place in the sky, and rocks fall through both air and water to return to their natural place, the Earth.

Because of this belief, Aristotle reasoned that heavy objects should fall faster than light objects. The heavier a rock, the greater its content of "earth," and therefore the stronger its tendency to return to the ground.

Events in everyday life seemed to support Aristotle's hypothesis—after all, a feather falls much more slowly than a rock.

For almost 2000 years, Aristotle's hypothesis was accepted as truth. Scientists did not experiment to test hypotheses, and there was little evidence to contradict his ideas. Scientific thought was governed only by logic and reason.

a) Drop a piece of paper and a book at the same time from the same height. Which one hits the ground first? Does your observation support Aristotle's hypothesis?

b) Can you think of an argument to disprove Aristotle's hypothesis?

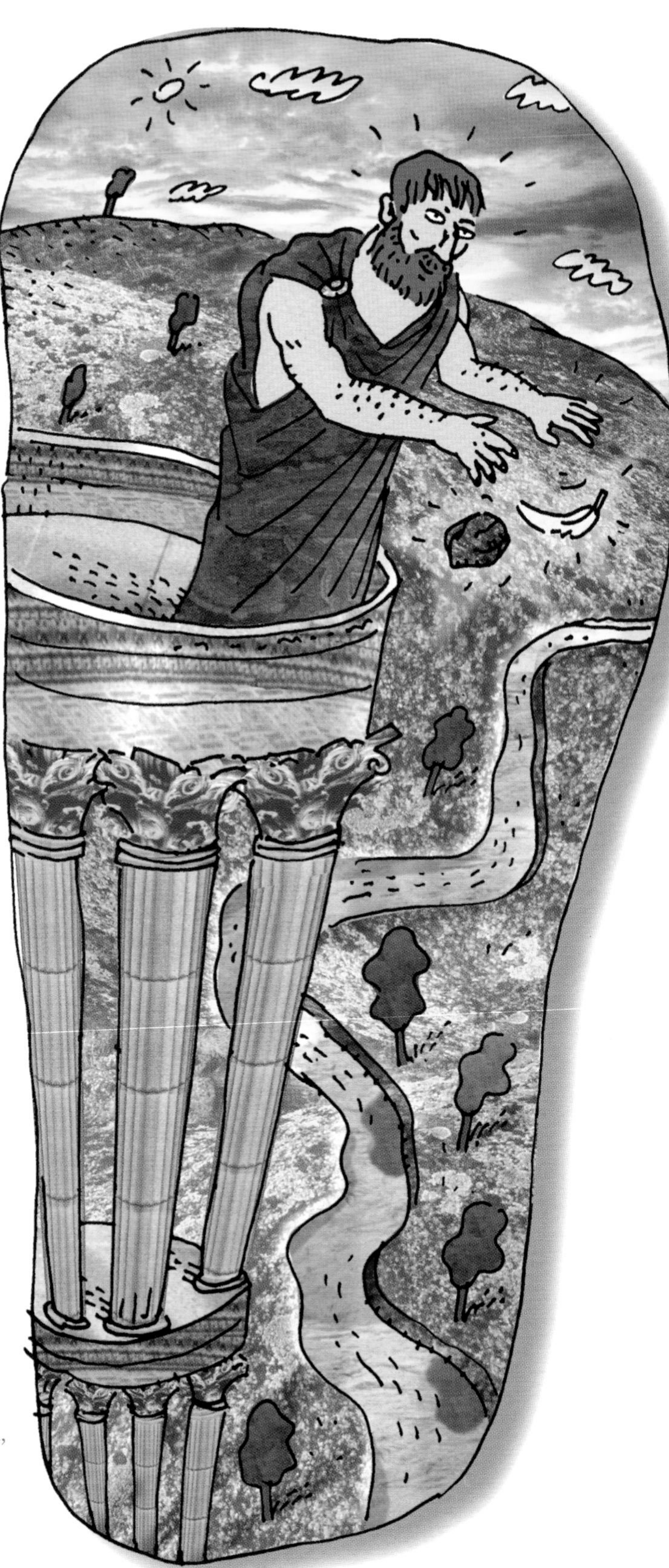

Which will land first, Aristotle's rock, or the feather?

## Galileo

Galileo (around A.D. 1600) thought he could defeat Aristotle's logic with logic of his own. Galileo imagined dropping two iron weights, a large one and a small one. According to Aristotle, the heavier weight should fall faster. Galileo then imagined the two weights connected by a rope.

c) If a large weight falls faster than a slow one, what should happen when Galileo drops the weights?

Galileo argued that if the smaller weight falls more slowly, it should slow the fall of the larger weight by pulling on the rope—the combined weights should fall slower than the large weight by itself.

But since both weights combined are heavier than the larger weight, the combined weights should fall faster. How can an object fall slower and faster at the same time? Galileo argued that heavy and light objects must actually fall at the same rate.

## Testing the Hypothesis

As times changed, so did science. By the time of Galileo, scientists saw the need to test their ideas. Having defeated Aristotle with logic, Galileo set out to prove his logic through experimentation.

d) Wad a piece of paper into a small ball and repeat the experiment of dropping paper and book at the same time. Which one falls faster this time? Was Aristotle or Galileo correct?

### Case Study Questions

1. Why do you think Aristotle's hypothesis wasn't questioned for so many years?
2. Besides gravity, what else did Galileo study? Try looking up his name in a CD-ROM or book encyclopedia.

### Extension

3. Design an experiment to test Galileo's hypothesis that all objects fall at the same rate.

# Energy

**E**NERGY IS THE ABILITY TO MAKE THINGS move. In other words, energy is the ability to do work.

You are using energy right now! Even if you are sitting perfectly still, your heart and other organs are working. They use energy from food you have eaten today to keep you going. You can see the words on this page because of light energy. Sound energy lets you hear the cars that are driving by. These are some of the forms of energy that are around you. The 10 common forms of energy are shown on these two pages. Can you think of another example for each one?

**electrical**

Static charges—electrical energy—can build up in the Earth and in clouds. When this energy is transferred, we see the small part of the electrical energy that is transformed into light (lightning).

**thermal**

The hot spray that rushes from the Earth in a geyser is created by thermal energy from deep in our planet.

**light**

Light energy from the Sun is transformed into many other forms of energy. For example, plants transform light into chemical energy.

**elastic**

As the arrow is released, the bow snaps back into its original shape. The "snap" is caused by stored elastic energy.

**gravitational**

The higher an object is above the Earth, the more gravitational energy it has.

**mechanical**

The car's fuel (stored chemical energy) is transformed into the mechanical energy of moving pistons, which powers the car.

## Transformation of Energy

Energy is important in its ability to change from one form to another through a process called **energy transformation.** Often, as one form of energy is transformed into another, the energy does work for us. The chemical energy in the food we eat is transformed by our bodies into mechanical energy when we move. When you ride in a car or a bus, the engine is transforming chemical energy in the fuel into mechanical energy that gets us to school on time.

## Conservation of Energy

Remember that energy can be changed from one form to another, but cannot be created or destroyed. This idea is called the **Law of Conservation of Energy.**

**sound**

Sound energy travels through substances, including the air.

**chemical**

When we eat, we stock up on chemical energy. Food is stored chemical energy.

**magnetic**

Moving charges create magnetic energy. We can harness this energy to do work.

**nuclear**

Energy stored in the smallest particles of matter can be released and transformed into electrical energy.

## SELF CHECK

1. Think about all the different ways you use energy. Get together with three other students. Use a metre stick to divide a large sheet of poster paper into six boxes. Use the instructions below to fill in the boxes. With your group, present your poster to the rest of the class.

**Poster Map**

| | |
|---|---|
| Print the names of at least 10 devices that use energy to do work. | Write a short story about how life on Earth would be different without energy to do work. |
| Draw a large picture of at least 3 of these devices. | Without looking back at the text, list as many of the forms of energy as you can. Draw a picture of each. |
| Draw a mind map of how you use energy every day. | List at least five energy transformations that you use every day. |

2. Make an acrostic for the word ENERGY. An acrostic is a series of words in which certain letters, when taken in order, spell out the word.

# *Force, Work, Energy, and the Roller Coaster*

FORCE, WORK, AND ENERGY COME TOGETHER in a very exciting place—the roller coaster ride at amusement parks. The ride operates using the scientific principles that you have been studying in this unit. Often physics students will visit local amusement parks to study the science involved in these rides. One question often arises—why would anyone invent such things? When did the rides first show up?

## The First Coaster

Russians built ice slides in St. Petersburg in the 15th century that were the precursors to modern roller coasters. A 22-m wooden frame was packed with snow and then watered down to create ice. Sand was added near the end of the run to stop the sleds, which were made from two-foot blocks of ice.

## The First Roller

La Marcus Thompson created the first roller coaster at Coney Island, New York, in 1884. His Scenic Railway was tame by today's standards. As the name implies, it was a way to ride along the beach and see the sights. The riders rode sideways to better see the ocean and rolled at the death-defying speed of 10 km per hour!

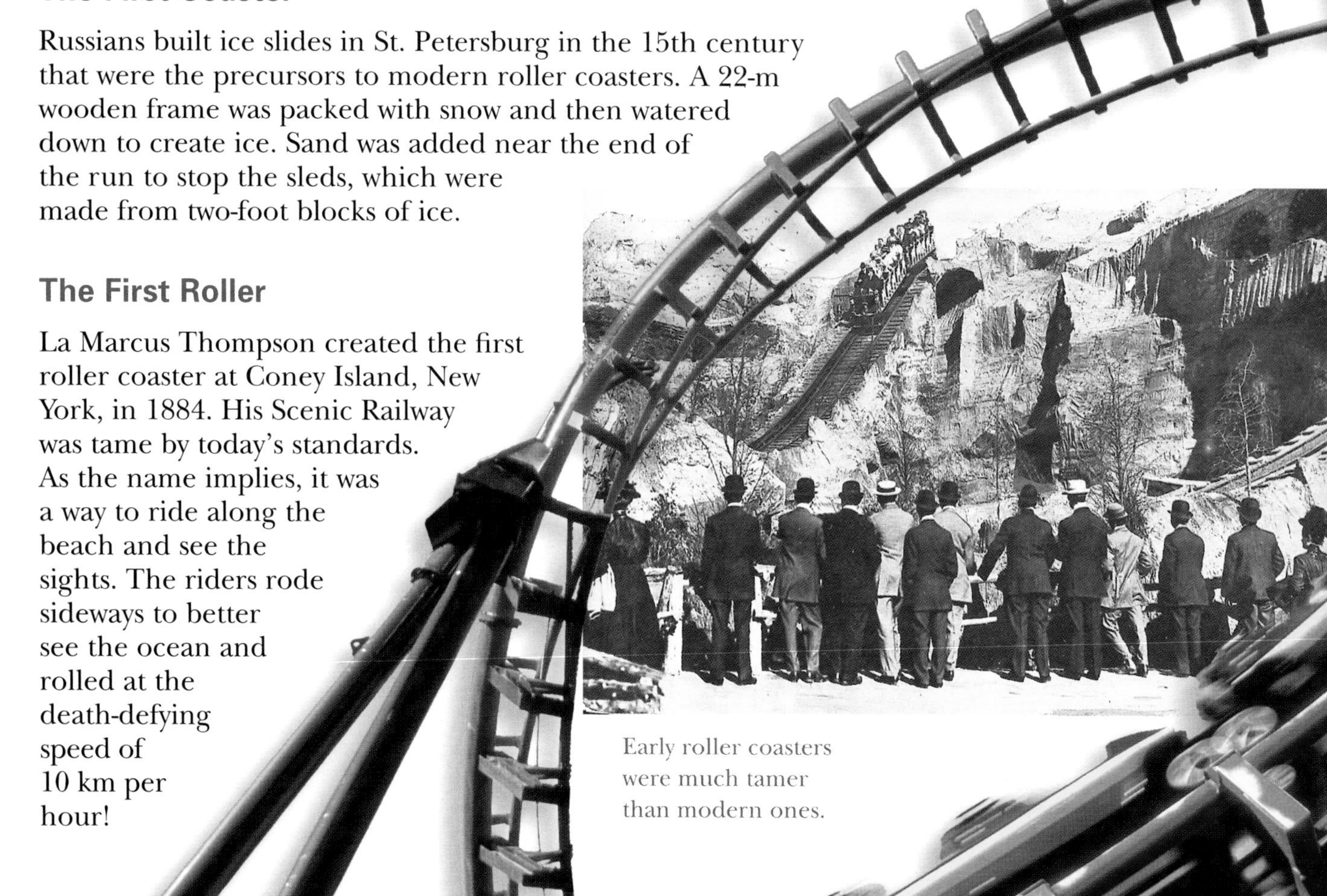

Early roller coasters were much tamer than modern ones.

## The First Thrill

Other inventors quickly entered the coaster market. The competition soon led to design improvements. The first oval-track coaster, which returned passengers non-stop to their starting point, and a chain elevator system that carried the loaded cars up the first hill sparked the development of bigger and faster roller coasters.

## Scary Coasters

The Roaring '20s ushered in the Golden Age of the roller coaster. One of the best was the Cyclone at Crystal Beach, Ontario. Perhaps the most ferocious coaster ever built, the Cyclone is the only roller coaster known to have had a nurse and first-aid station at the exit for its passengers!

## Disney Saves the Ride

After the Depression and World War II, amusement parks were saved, in part, by the creation of the theme park, an idea of Walt Disney's. Initially people scoffed at Disney, but he struck an unusual deal that tied his theme park to newly created television shows, the Mickey Mouse Club and The Wonderful World of Disney. Disneyland and Disney's television shows both took off in popularity, and amusement parks have never been the same since.

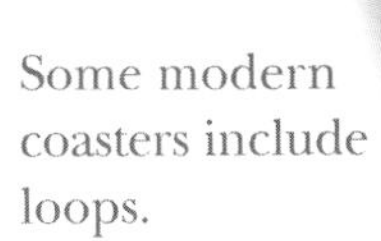

Some modern coasters include loops.

### SELF CHECK

1. Create a time line to represent the history of the roller coaster.
2. Create a poster to advertise a "new, exciting" roller coaster ride at a local amusement park.
3. Design a new roller coaster ride. What would you add to make it a thrilling experience?

### Apply

4. a) What forces act on you during a roller coaster ride? How do these forces add to the thrills that you experience?

   b) When is work done on the roller coaster cars and their passengers?

5. What simple machines can you find on a roller coaster?

# *Keeping the Rides Safe*

WHO DESIGNS AND BUILDS AMUSEMENT PARK RIDES? Who keeps them in top condition and safe for the millions of passengers that ride them?

Meet Norm Pirtovshek. His job is to make sure that the rides at Canada's Wonderland, near Maple, Ontario, are safe. To do that, he needs to know all about forces, work, and energy. The roller coaster cars loaded with people put stresses and strains on the cars, on the tracks, and on the support structure of the ride. Each ride is checked daily for abnormal wear and is thoroughly inspected in the off-season. Metal track is X-rayed. This allows the maintenance workers to see previously invisible cracks in the metal. These cracks, caused by the forces created by the rides, can be dangerous if not repaired.

Wheels are lubricated and bearings checked to make sure there is no unnecessary friction to slow the ride. Remember, in a roller coaster, a chain drive pulls you to the top of the first hill and gravity gives you the rest of the ride for free. With too much friction on the ride, you may not be able to get back to the starting point.

## Let's Go!

The roller coaster was one of the first amusement park rides invented. They were originally made of wood and the coaster cars rode on steel wheels. Later versions followed paths of steel and rolled on air-filled tires. Basically, they all work for the same reason: the force of gravity. The park uses a machine, which does work to pull you to the top of the first hill. The rest of the time gravity does the work.

## TRY THIS

### Build a Coaster

Design your own amusement park ride.

Use Hot Wheels track and cars to design and build a kind of roller coaster. What do you have to do in order to make sure that the cars travel the entire length of the track without flying off?

Use the picture on these pages, your experience, and what you have learned about roller coasters to answer the following questions.

**a)** What happens to the size of the hills through the ride? Why do you think they are built that way?

**b)** Are you moving faster or slower when you are at the top of a hill? Why?

**c)** Are you moving faster or slower when you are at the bottom of a hill? Why?

**d)** When you are in a turn, are you pushed inward or outward? What causes this force?

**e)** Are the tracks flat or are they tilted inward or outward on curves? Why are they designed this way?

19

# The Science of Toys

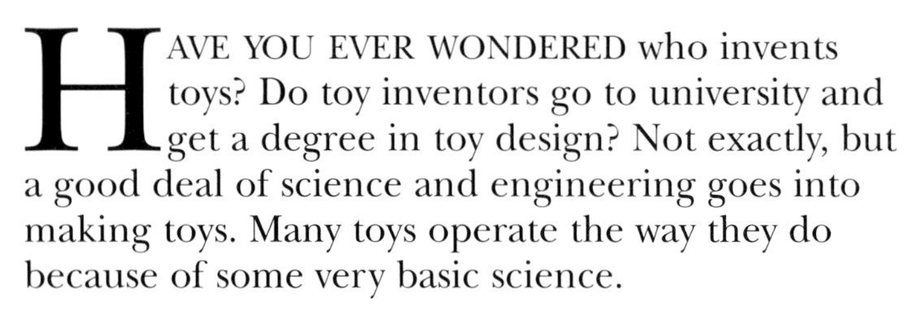

HAVE YOU EVER WONDERED who invents toys? Do toy inventors go to university and get a degree in toy design? Not exactly, but a good deal of science and engineering goes into making toys. Many toys operate the way they do because of some very basic science.

## Early Moving Toys

In the 18th and 19th centuries, people were amazed by moving toys called automatons that seemed almost human. One such automaton was The Writer, a child-like doll that could actually write a 50-word letter. Many people in these times didn't understand this new kind of toy. Some thought what they were seeing was a miracle or witchcraft.

In 1887, the American inventor Thomas Edison developed the first speaking toy, by fitting a doll with a miniaturized phonograph and recorded disc.

## Hidden Works

The mechanisms that make toys move are generally hidden from view, making them all the more mysterious. But just what does go inside mechanical or other kinds of toys? In the case of the duck, a wound spring releases stored energy to drive a gear train.

Another component, called a cam, converts the rotating motion of the gear train to back-and-forth motion in a lever. This lever drives the duck's feet, and off it scoots!

## Modern Toys

Today's toys may seem much more complex and perhaps more exciting than the old-fashioned ones, but the mechanisms inside moving toys haven't changed very much in the past 400 years. Many modern toys have sophisticated electronic components, but most mechanical toys still rely on the levers, gears, and pulleys that have been powering toys and delighting children for decades.

### SELF CHECK

**1.** Bring in several toys from home and study the science of how they work.

**a)** Wind-up toys.

See if you can discover what makes each one "go." Try identifying the components and explain with words and sketches how these components work together to make the toy move. Do they work as well on different surfaces—say, a wooden desk and a carpeted floor?

**b)** The yo-yo.

Why does it climb back up its string? What forces are at work as the yo-yo does down and then back up? How do you make a yo-yo "sleep"?

**c)** The gyroscope.

Why does a spinning gyroscope stay spinning in one place? What happens as the spinning motion slows down? What forces are at work in a spinning gyroscope?

**d)** Other toys, like Pogs, Meccano, or Hot Wheels cars.

What scientific principles do they use?

20

SCIENCE AND LITERATURE

# Science and Fiction

THE FUNDAMENTALS OF SCIENCE are sometimes used in movies, television, and book plots. Characters from James Bond to Nancy Drew have used resourcefulness, ingenuity, and scientific knowledge to solve problems and escape dangerous situations.

One group of scientific principles that are often used are those involving force, work, and energy. James Bond's Aston Martin, with all its special gadgets, and MacGyver's ingenious devices made from ordinary objects, use these principles.

You can use all the knowledge that you have gained in this unit to write your own action/adventure story.

## Story 1

"Let him go!," shouted Nancy. "Whatever you've got against me, he's not involved."

Murphy smiled, "Yes. But you said yourself that sometimes innocent people get caught in the crossfire." With that, Murphy closed the door.

Nancy listened to the lock being set and Murphy walk away before she peered out the small window in the door. In the next room, her grandfather struggled against his rope bonds, to no avail. On the table beside him, the timer on Murphy's incendiary bomb ticked toward zero. 29:32, and counting.

Nancy knew she had to get out. She turned and studied the room.

**(You write this part!)**

With two seconds left, Nancy pulled the wires from the detonator and the lights on the timer went dark.

### Story 2

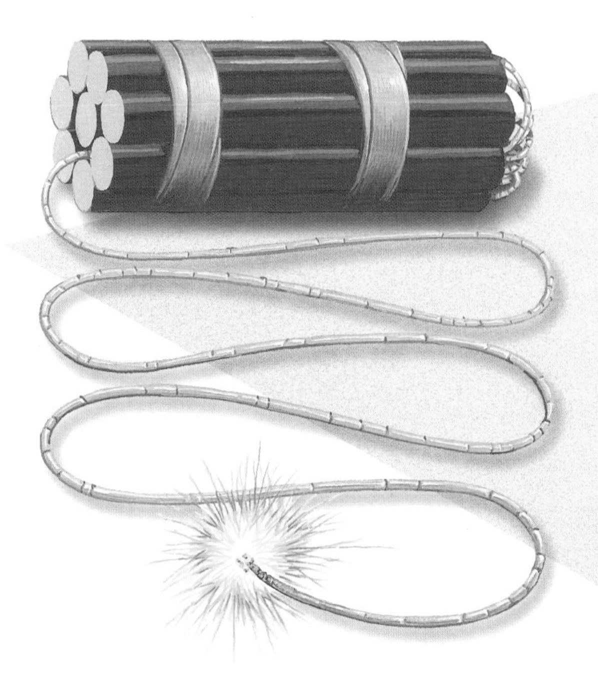

Drago struggled against the chains that held him. Turning to Pete he said, "It's no use. They won't budge."

"Murdoch really thought this one out," Pete replied. "No matter how hard we try, we are still two metres away from the fuse."

Looking across the room, Pete could see the fuse burn slowly toward the dynamite. "If only we could make some kind of device that would move two metres and then put out the fuse."

Drago looked around the room. "Maybe there is a way."

Stretching as far as he could, he grabbed a box. After surveying its contents he said, "See if you can reach that other box."

Pete hooked the box with his foot and kicked it back. "What do you have in mind?" he asked.

"Just watch," replied Drago.

**(You write this part!)**

The fuse was out. Pete and Drago were saved.

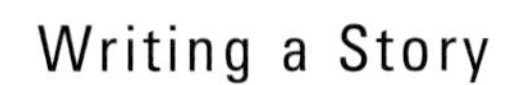

## TRY THIS

### Writing a Story

- Use either of the two story starters above. It will set the general scene for you. You can add details.
- Work individually for a few minutes to generate ideas for the story plot.
- Next, work with a partner. Share your ideas. Using ideas you generate, work together to develop one story.
- With your partner, write the story. You must start with the first lines you are given and end with the last line.
- After writing the story, exchange stories with another pair of writers. Each pair can then offer constructive advice about the stories of the other. Using this editorial advice, write a second draft of the story.

**a)** What science did you use in your story? How did you use force, work, and energy in the plot?

**b)** Was the science you used accurate? Will the readers have to stretch their imagination to believe your story?

**c)** In each case, the location was not specific. What kind of a room did you create for the hero or heroine to have the materials necessary to escape?

**d)** Draw a series of pictures that show the action in your story. In movies and television, this is called making a story board. It is very useful to producers and directors.

**e)** Watch cartoon shows on television. Do you see situations where science is disregarded? (For example, when Wile E. Coyote runs off a cliff, gravity only works when he looks down.) Make a note of some of these situations.

## Key Outcomes

Now that you have completed this unit, can you do the following? If not, review the sections listed in brackets.

Identify the different types of forces. (1)

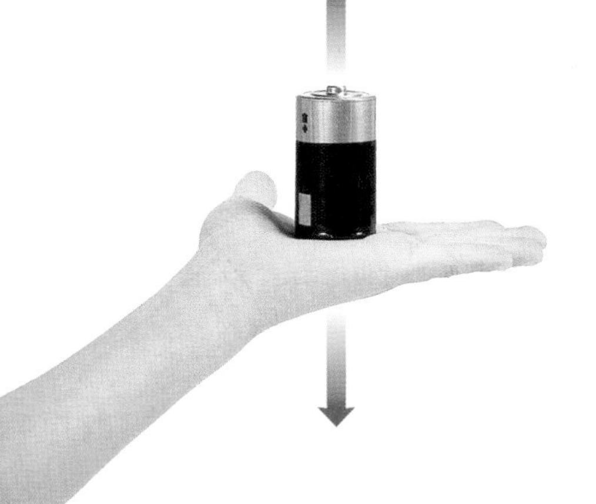

Differentiate between balanced and unbalanced forces. (2)

Describe the effects friction and buoyancy have on movement.(2, 6, 8)

Explain how forces on objects cause movement. (3)

Interpret graphs that show the relationship between the stretch of an elastic band and the force applied to it. (4)

Describe a force meter and explain its limitations. (5)

Identify careers that use a knowledge of force, work, and energy. (9, 18)

Describe how forces in the Earth cause earthquakes and discuss what cities can do to lessen their effects. (9)

Use the formula $W = F \times d$ and direct measurements to compute the amount of work done. (11)

Recognize that a machine is simply a device that converts one force (or form of energy) to another. (12, 13, 14)

Identify some of the advantages and disadvantages of common machines. (12, 14)

Explain how a ramp reduces the effort needed to raise a load, but not the work done. (13)

Describe the differences between the hypotheses of Aristotle and Galileo on the force of gravity. (15)

Describe the forms of energy. (16)

Explain how forces and energy are used in design. (17, 19)

Use creative writing to express how force, work, and energy could be used in an adventure story. (20)

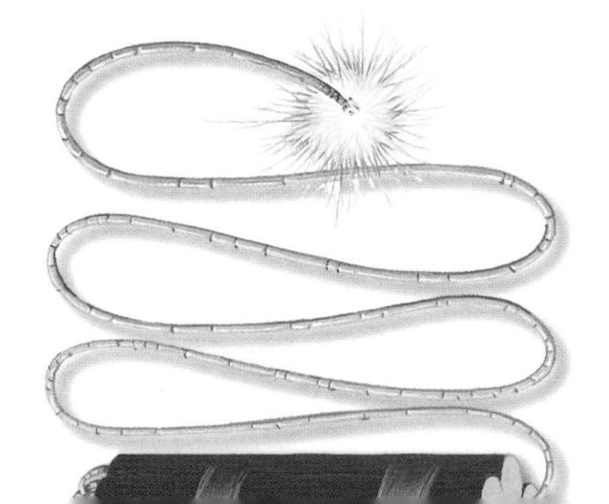

## Review Questions

You can assess how much you have learned about force, work, and energy by trying these questions. Write the answers to the questions in your notebook.

1. Draw a diagram of a hand holding a helium-filled balloon. Draw in the forces acting on the balloon, and label the forces. Are the forces balanced or unbalanced? When the hand lets go of the balloon, are the forces balanced or unbalanced?

2. Note whether the following statements are true or false. If a statement is false, rewrite the statement so that it is true.

   **a)** A force is something that pushes an object.

   **b)** When two forces act on an object in opposite directions and are equal, the forces are balanced.

   **c)** Friction is a force that resists motion.

   **d)** The strength of friction will vary depending on the roughness or smoothness of the surfaces that come into contact.

   **e)** Newton's First Law of Motion states that for every action force, there is an equal and opposite reaction force.

   **f)** The unit of measurement for force is the joule.

   **g)** The unit of measurement for work is the newton.

   **h)** The screw is a simple machine similar to an inclined plane and can multiply force.

3. The baseball bat in the photograph at top right is a lever. Where is

   **a)** the fulcrum?

   **b)** the effort force?

   **c)** the load force?

4. Copy the following sentences, filling in each blank with the word or phrase that correctly completes the sentence. Use words from the following list: balanced forces, distance, energy, energy transformation, friction, force, lever, gravity, inclined plane, mechanical advantage, pulley, screw, simple machine, unbalanced forces, wedge, wheel-and-axle, work. (You will not need to use all of the words.)

   **a)** ________ is a force that resists motion whenever one materials rubs against the surface of another.

   **b)** If a ________________ acts on an object, it will cause a change in the motion of the object.

   **c)** The inclined plane, lever, screw, wedge, pulley, and wheel-and-axle are all examples of a ________________.

   **d)** Work is equal to ____________ times ____________.

   **e)** A book sitting on a table has ________________ acting on it. One of these is ________________.

   **f)** Energy is changed from one form to another in a process called ________________.

   **g)** The amount that a machine can multiply force is called ____________.

5. List the places where you see a pushing force. What is doing the pushing?
6. List the places where you see a pulling force. What is doing the pulling?
7. List 3 places where you see the force of gravity at work.
8. List 3 places where you think the force of friction is at work.
9. List 3 places where springiness or elasticity is at work.
10. List 3 places where unbalanced forces are at work. Choose one of them, draw a diagram of it, and place arrows on the diagram representing the forces.

Examine this picture, then answer questions 5 to 10.

11. a) The knife is a machine. What kind of simple machine is it?

    b) Which simple machine is most like a see-saw?

12. Copy the letter of each description of a form of energy in your notebook. Beside each letter, write the word from the right column that best fits the description.

    a) the form of energy you can see

    b) the energy that an object has when it is above the surface of the Earth

    c) the energy of moving particles in an object

    d) the form of energy that causes some kinds of metal to attract or repel some metal objects

    e) the energy stored in the central part of an atom

    f) the energy of moving electrical particles

    g) the energy stored in a object when its shape is changed by stretching or compressing

    h) the form of energy that you can hear

    i) the form of energy stored in matter

    j) the energy in any set of moving parts

| |
|---|
| mechanical |
| elastic |
| thermal |
| light |
| magnetic |
| electrical |
| gravitational |
| nuclear |
| sound |
| chemical |

## Problem-Solving

13. If you place a force of 12 N on a rubber band, it stretches 8 cm. If you then double the force you place on the rubber band, how much will it now be stretched?

14. If you measure the stretched length of a rubber band as varying forces are added, what will the graph of force vs. length look like?

15. A force of 5 N moves an object 10 m. How much work is done on the object?

16. A pulley system allows you to lift a 1000-N load with a force of only 250 N. This seems to suggest that you do less work to lift the object. Is this true? Explain.

17. It takes 100 N to push a box up an inclined plane. To lift the box that same height would require 400 N. What is the mechanical advantage of the inclined plane?

## Challenge

18. Explain the statement "Friction is both useful and a problem."

19. Machines have helped society by reducing the need for human labour. What are the benefits and problems that these machines have created for society?

20. Look at the photograph below of the Canadarm on a U.S. space shuttle. Think of at least two simple machines that must form a part of this device, enabling it to move and to pick up objects.

# Glossary

**A**

**air resistance:** friction caused by air (p. 16)

**B**

**balanced forces:** forces of equal strength that act on an object in opposite directions (p. 6)

**E**

**effort force:** the force needed to move an object (p. 27)

**energy:** the ability to make things move—the ability to do work (p. 34)

**energy transformation:** when energy changes from one form to another (p. 35)

**F**

**force:** a push or pull on an object; gravity, friction, magnetism, electrostatics, and buoyancy are examples (p. 6)

**friction:** a force that resists motion whenever one material rubs against another (p. 7)

**J**

**joule (J):** the unit used to measure work and also energy; one joule is the amount of work done when a force of one newton pushes or pulls an object a distance of one metre (p. 25)

**L**

**Law of Conservation of Energy:** energy can be changed from one form to another, but cannot be created or destroyed (p. 35)

**load force:** the force needed to move an object without a machine (p. 27)

**lubricant:** a substance that reduces friction and wear on moving parts (p. 16)

**M**

**machine:** a device that helps people do work more easily; machines may change the direction of forces, transfer forces from one location to another, multiply force, speed, or distance (p. 26)

**mechanical advantage:** the amount that a machine multiplies a force, measured by dividing the load force by the effort force (p. 27)

**N**

**newton (N):** the unit used to measure force (p. 7)

**Newton's three Laws of Motion:** the summary of Isaac Newton's theories about motion (p. 8)

**Newton's First Law:** An object at rest will stay at rest and an object in motion will stay in motion, unless acted on by an outside force. (p. 8)

**Newton's Second Law:** An object will move with an acceleration—increasing speed—proportional to the force applied to it. (p. 8)

**Newton's Third Law:** For every action force, there is an equal and opposite reaction force. (p. 8)

**S**

**simple machines:** the six fundamental machines that are combined to make all other machines; the wedge, the screw, the lever, the wheel-and-axle, the pulley, and the inclined plane (p. 27)

**U**

**unbalanced forces:** result when one force acting on an object is greater than another force acting in the opposite direction (p.7)

**W**

**work:** if a force has been exerted on an object and the object moves for a distance in the direction of the force, then work has been done. The relationship between work, force, and distance is described in an equation:

$W = F \times d$ (p. 24)